T0140585

Gaming Media and Social Effects

The scope of this book series is inter-disciplinary and it covers the technical aspect of gaming (software and hardware) and its social effects (sociological and psychological). This book series serves as a quick platform for publishing top-quality books on emerging or hot topics in gaming and its social effects. The series is also targeted at different levels of exposition, ranging from introductory tutorial to advanced research topics, depending on the objectives of the authors.

More information about this series at http://www.springer.com/series/11864

Yiyu Cai · Wouter van Joolingen
Zachary Walker
Editors

VR, Simulations and Serious Games for Education

Editors
Yiyu Cai
Nanyang Technological University
Singapore, Singapore

Wouter van Joolingen
Freudenthal Institute
Utrecht University
Utrecht, The Netherlands

Zachary Walker
Institute of Education
University College London
London, UK

ISSN 2197-9685 ISSN 2197-9693 (electronic)
Gaming Media and Social Effects
ISBN 978-981-13-4810-5 ISBN 978-981-13-2844-2 (eBook)
https://doi.org/10.1007/978-981-13-2844-2

This Springer imprint is published by the registered company Springer Nature Singapore Pte Ltd.
The registered company address is: 152 Beach Road, #21-01/04 Gateway East, Singapore 189721, Singapore

Contents

Introduction

Yiyu Cai, Wouter van Joolingen and Zachary Walker

Abstract This book is our continuous effort (Cai in 3D Immersive and interactive learning. Springer, Berlin, 2012; Cai and Goei in Simulation, serious games and their applications. Springer, Berlin, 2014; Cai et al. in Simulation and serious games in education. Springer, Berlin, 2016) to promote simulation and serious gaming. The eleven chapters book presents a multi-facet view of simulation and serious games for educational applications from STEM to Special Needs. Virtual Reality is one of the emphases in this book.

1 Background

In 2012, the first Asia-Europe Symposium on Simulation and Serious Games was held at Nanyang Technological University, Singapore. Best papers selected from the symposium were published in a book Simulation, Serious Games and Their Applications by Springer (Cai and Goei 2014). The second version of Europe-Asia Symposium on Simulation and Serious Games was held at Windesheim University of Applied Sciences, The Netherland, in 2014. A book entitled Simulation and Serious Games in Education based on the selected papers from the 2014 symposium was published by Springer (Cai et al. 2016). The third version of Asia-Europe Symposium on Simulation and Serious Games was held in Beijing Normal University at Zuhai in 2016 as part of the 2016 ACM SIGGRAPH International Conference on

Y. Cai (✉)
Nanyang Technological University, Singapore, Singapore
e-mail: myycai@ntu.edu.sg

W. van Joolingen
Freudenthal Institute for Science and Mathematics Education,
Utrecht University, Utrecht, The Netherlands
e-mail: w.r.vanjoolingen@uu.nl

Z. Walker
Institute of Education, University College London, London, UK
e-mail: zacharywalker0@gmail.com

Y. Cai et al. (eds.), *VR, Simulations and Serious Games for Education*, Gaming Media and Social Effects, https://doi.org/10.1007/978-981-13-2844-2_1

Fig. 1 The 2016 Asia-Europe Symposium on Simulation and Serious Games as part of the ACM SIGGRAPH VRCAI 2016 Conference

Virtual-Reality Continuum and Applications in Industry (VRCAI 2016) (see, Fig. 1) (Cai and Thalmann 2016). Partially based on the 2016 symposium, this new book is devoted to Virtual Reality, Simulation and Serious Games in Education. For those best papers selected from the symposium presentation, substantial enhancements are made before they are accepted as book chapters in the new book.

2 About the Book

This book has eleven chapters organized as follows.

This chapter is an introduction by the book editors Yiyu Cai, Wouter van Joolingen and Zachary Walker. In Chap. 2, Veermans and Jaakkola will share their work on design considerations for educational simulations and games. In Chap. 3, Anne van der Linden and Wouter van Joolingen will present their work using a serious game supporting conceptual change in mechanics. In Chap. 4, Casano et al. will describe the evaluation of a re-designed framework for embodied cognition math games. In Chap. 5, Ryan et al. will elaborate their research on the use of virtual & augmented reality technology to enhance the learning and understanding of biological molecules. In Chap. 6, Wu and Zheng will discuss their study on autism education through motion sensing based gaming. In Chap. 7, Yang et al. will investigate vehicle behaviours simulation technology based on neural network. In Chap. 8, Liang et al. will showcase their multi-player, and cross-platform competitive social game BlockTower. In Chap. 9, Hovardas and Zacharia will discuss an inquiry-based approach for learning system dynamics and modeling of the prey-predator system. In Chap. 10, Siti et al. will report their research and development on VR Serious Game for Special Needs Education. In Chap. 11, Xie et al. will describe Virtual Reality Simulation for Engine Disassembly with Natural Hand-Based Interaction.

Researchers and developers in Simulation and Serious Games for educational use will benefit from this book. Training professionals and educators can also benefit from this book by learning the possible applications of Virtual Reality, Simulation and Serious Games in various areas.

References

Cai, Y., & Thalmann, D. (2016). The 3rd Asia-Europe symposium on simulation & serious gaming. In *Proceeding of the ACM SIGGAPH Conference on Virtual-Reality Continuum and Applications in Industry*, Zhuhai, China, December 03–04, 2016.

Cai, Y. (Ed.). (2012). *3D immersive and interactive learning*. Berlin: Springer.

Cai, Y., & Goei, S. L. (Eds.). (2014). *Simulation, serious games and their applications*. Berlin: Springer.

Cai, Y., Goei, S. L., & Trooster, W. (Eds.). (2016). *Simulation, serious games in education*. Berlin: Springer.

Pedagogy in Educational Simulations and Games

Koen Veermans and Tomi Jaakkola

Abstract Educational simulations and serious games hold great potential for creating engaging and productive learning environments in science, technology, engineering and mathematics (STEM) domains. In this paper, we present and reflect on some of our research findings from a series of studies on a computer simulation in the domain of electricity. These studies used the same simulation with varying instructional designs and over a range of grades. Interestingly, each design had a unique influence on either student performance or student engagement, or both. We hope our results can provide insight for designers producing simulations (or, serious games) for education and for educators utilizing these designs in practical settings.

1 Introduction

From their inception, educational simulations have held the promise of creating engaging and productive learning environments in science, technology, engineering and mathematics (STEM) domains. Some of the advantages that have been put forward in the literature include simulations being learner-centric, scalable, reusable; having affordances related to illustration and visualization; leading to student interest and engagement; and producing desirable learning outcomes, particularly in terms of conceptual knowledge but also with regard to developing understanding about scientific inquiry (Slavin et al. 2014).

In addition to these advantages, the current learning analytics trend towards obtaining learner data in order to analyze productive learner behavior also adds to renewed and strengthened interest in educational simulations and serious games. However, this trend does not mean that the outcomes of learning with and from educational simulations or games are straightforward or always positive. In this paper, we will

K. Veermans (✉) · T. Jaakkola
University of Turku, Turku, Finland
e-mail: koen.veermans@utu.fi

T. Jaakkola
e-mail: tomi.jaakkola@utu.fi

Y. Cai et al. (eds.), *VR, Simulations and Serious Games for Education*, Gaming Media and Social Effects, https://doi.org/10.1007/978-981-13-2844-2_2

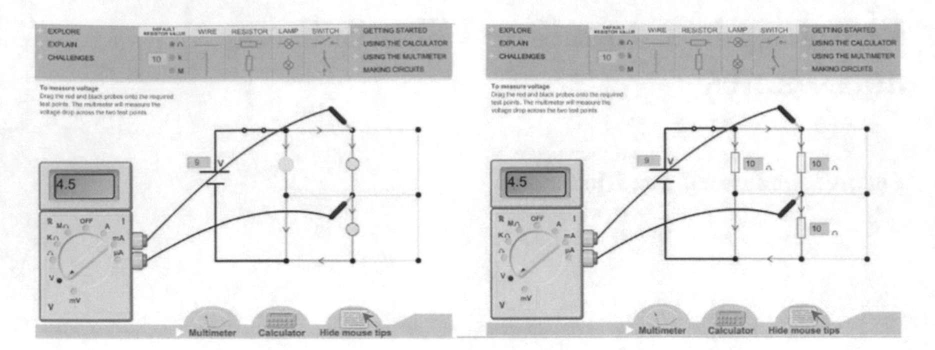

Fig. 1 The simulation with bulbs (left) and resistors (right) as used in the studies

present and reflect on some of our research findings from a series of studies with a computer simulation in the domain of electricity (e.g. Jaakkola and Nurmi 2008; Jaakkola et al. 2010, 2011; Jaakkola 2012; Jaakkola and Veermans 2015, 2018). Through these studies, our aim is to demonstrate that sound pedagogical design can make a simulation (or a game) suitable and effective across a wide age range of pupils, and that different pedagogical decisions can have notable impact on students' learning.

2 General Settings in the Studies

In the studies that are reviewed in this paper, student participants used the same simulation (see Fig. 1) to build circuits, observe circuit behavior, and study the properties and underlying principles of electric circuits. The representation level of the simulation was semi-realistic; it displayed circuits schematically but also included light bulbs with dynamically-changing brightness and realistic measuring devices. The simulated model was authentic but for two exceptions: The wires had no resistance and the battery was always ideal (i.e. there was no change in its potential difference with time).

The students' inquiry process with the circuit simulation was supported and guided by instructional worksheets designed to confront and overcome common misconceptions about electric circuits. In general, the worksheets asked students to construct various circuits and conduct electrical measurements with the simulation. The worksheet also provided scaffolding for the students to predict, investigate and infer how the changes and differences in circuit configurations affected circuit behavior. The worksheets began with a very simple and structured task, wherein the students were asked to construct a circuit with one battery, wires, and a bulb. Subsequent tasks were progressively more challenging and open-ended, requiring students to construct more complex circuits that met a specific criterion (e.g. brightness of bulbs has to be $A > B = C$).

The general procedure was identical across all studies. In the beginning, students took a pre-test designed to assess their baseline knowledge of electric circuits. The pre-test scores were then used to assign students into the different conditions. Matched pairs of students were created based on the pre-test scores, and the students in each pair were allocated randomly to either of the conditions. This procedure ensured relatively small differences in pre-test knowledge between the conditions, which made the assessment of learning gains during the intervention easier between the conditions. After students were allocated into the conditions, random pairs of students were created within conditions. These pairs then had approximately 90 min to build and study circuits in the simulation and solve various circuit challenges listed in the worksheets. To assess students' level of engagement during the simulation task, students were asked to indicate their situational interest in the beginning, middle, and end of the intervention in some of the studies. The post-test that was designed to assess changes in students' subject knowledge during the intervention was administered one day after the intervention. Although the students worked in pairs during the intervention, they completed all the tests individually.

Interestingly, though the overall impact was predominantly positive, each design had a unique influence on student performance and/or engagement. We hope our results can provide some new insights for designers when designing simulations (or, serious games) for education and for educators utilizing these designs in practical settings.

3 Learning Outcomes, Interest and Learning Time

The general goal of our studies has always been to study learning outcomes across different settings, but gradually, due to reports and literature indicating that students' motivation and interest towards science start declining from their early years at school (European Commission 2011; Osborne and Dillon 2008; Vedder-Weiss and Fortus 2011, 2012), interest in science (more specifically, interest during science tasks) became an integral part of our investigations (Tapola et al. 2013, 2014). In other words, the goal should be to design learning environments that are both productive (good learning outcomes) and engaging (motivating from students' perspectives).

As a starting point, we present the global findings on learning outcomes and interest that were obtained in studies across a range of grades (Fig. 2) As can be seen from the first graph, regardless of their initial knowledge, students gained knowledge while interacting with the simulation-based learning environment in all five grades, with the smallest gain occurring in grade 4, the largest in grade 5, and similar gains in grades 6 and 8 (the overall outcome shows a significant linear interaction between grade level and post-test scores; students scored higher as a function of grade level, $p < .001$). Interest was above midpoint for all grades, but the second graph shows the presence of a tendency for decreasing interest with increasing grade.

Based on the overall outcomes, it can be argued that the learning environments used in these studies fulfilled the aims of the learning environments both in terms

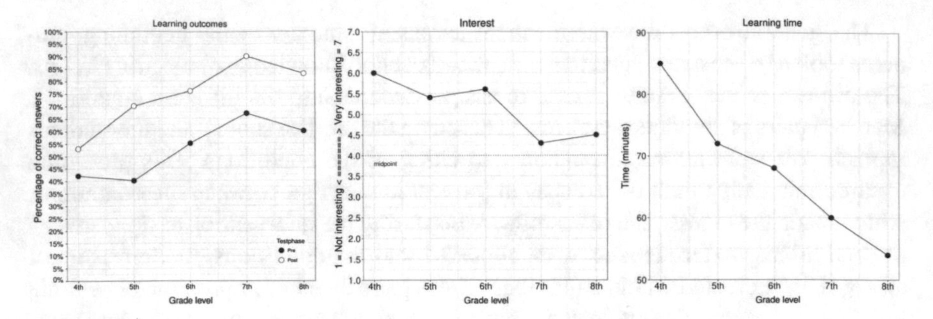

Fig. 2 Trends in learning outcomes (pre- and post-), interest and learning time across grades 4–8 (9–15 years)

of learning outcomes and interest. However, the results also showed that different outcomes regarding learning and interest were obtained over the different grades. This was the first indication that simulation design was not the only factor that affected participants' learning outcomes and experiences. More indications will be pointed out in the following sections in which we investigate the impacts that different design elements and decisions had on the outcomes in greater detail.

4 Instructional Support

In the previous section, we argued that it was not the simulation alone that mattered, since the results showed clear differences across grades. Learner support in the instruction was more structured in the beginning but became gradually less structured and more open-ended towards the end. The results of the 7th and 8th graders in terms of learning process, outcomes and interest levels supported the idea that this shift from structured to open-ended lay at the core of the successful use of the same simulation learning environment over a range of grades. The 7th and 8th graders were quick (about 20% less time needed to complete the learning tasks than elementary school students), but not highly interested during the early phases of the learning task with relatively structured instruction. Their perceptions changed when the tasks became more open-ended, resulting in higher perceived interest towards the end (interest at time $1 = 4.1$ while interest at time $3 = 4.5$) and good learning outcomes ($p < .001$, see Fig. 2). This showed that an environment with instructional support that built on the idea of fading from structured to open-ended was flexible towards individual differences among learners and could thus address students with different proficiency levels within a grade, but also provide learners across a wider age range with engaging and productive learning experiences.

For younger learners, it was not only about instruction being more or less structured (the 4th graders' learning outcomes might indicate that the more open-ended tasks at the end were too difficult for them) but also about how that structure was presented, as was illustrated by the results from a study that varied the type of instruction in terms of the level of scaffolding (Jaakkola et al. 2011). One condition received implicit instruction—procedural guidance (e.g. what kind of circuit to construct and what to measure)—while the other condition received more explicit instruction in which the implicit instruction was accompanied with the rationale behind the explorations (e.g. students received guidance on where to focus their attention). The results showed clear differences between the two conditions both in terms of learning outcomes (higher in the explicit condition: post = 11.29, pre-post $p = .013$, d = .78, vs. post = 9.75, pre-post $p = .31$, d = .22) and in terms of learning time (longer in the explicit condition: 79 min. vs. 66 min.). This result also highlighted that it was not just about the simulation, but also about the pedagogy and what they together triggered in the learner. For learning to occur, a learner should not just be interacting with a learning environment but also be cognitively engaged in the activity. Both the shorter learning time and the lower learning gains suggested that this was not happening in the condition with implicit instruction.

5 Perceptual Concreteness of Simulation Elements

An important but often overlooked design consideration for the development of educational simulations is the determination of how concrete the simulation (elements) should be, since the perceptual and conceptual concreteness of any representation can greatly affect what the students learn and how they can utilize that knowledge in other situations. A simulation with concrete elements is easier to relate to and understand, but extensive amount of detail can also hinder extraction of relevant information and result in overly contextualized understanding. More abstract elements in a simulation can highlight relevant information and support generalization, but at the cost of becoming more difficult to understand. As a result, concreteness fading–starting with concrete elements to ensure proper contextualization, then switching to abstract elements to ensure that understanding is less bound to specific contextual details—has been proposed as an optimal solution (Fyfe et al. 2014).

In the studies for which the outcomes were summarized above, learners either used a simulation environment that employed this idea of concreteness fading or used a simulation with solely concrete elements. In the concrete condition, learners used concrete elements (bulbs) throughout the learning phase while in the fading condition, they started with concrete elements (bulbs) but switched to abstract elements (resistors) after some time. Though bulbs and resistors could theoretically be argued to be equally concrete or abstract (equivalent in the physics sense—bulbs are a special case of resistor), our empirical evidence shows that from a learner's point of view (i.e. perceptually and conceptually) they are not. In the pre-test, questions related to bulbs were answered correctly significantly more often than questions on resistors (both 5th and 6th and 7th and 8th, $p < .01$). Moreover, the first study with this design comparison of the two conditions in terms of learning outcomes showed that learners (5th and 6th graders) in the concrete condition outperformed learners in the fading condition on the post-test ($p < .05$, consistent over both grade levels; Jaakkola and Veermans 2015). As it turned out, a considerable number of the 5th graders in the fading condition failed to complete the learning tasks within the given time frame, indicating a clear difficulty compared to the concrete condition where practically all participants were able to complete the tasks within the given time frame. In the fading condition, 11% of the 5th graders and 75% of the 6th graders completed all tasks while in the concrete condition, 75% of the 5th graders and 100% of the 6th graders completed all tasks (χ^2 for all $p < .001$). This result indicated that students' interaction with resistors rather than bulbs caused the learning process rate and learning outcomes to deteriorate. During this study, perceived interest was also assessed at several points in time, with an in-depth analysis of interest (Tapola et al. 2013) showing an interaction between the level of interest for students and condition ($p = .002$, $\eta^2 = .13$; interest in the concrete condition increased while interest in the fading condition decreased during the conditions' interactions with the learning environment), revealing that concrete elements were to be preferred for these students from an interest perspective at this age, in addition to being better from a learning perspective as found in Jaakkola and Veermans (2015).

In a follow-up study, the fading from concrete to abstract elements was slightly delayed (i.e. students used bulbs a bit longer before moving to resistors), with the underlying hypothesis that the delay might help students align and link the representations better (Jaakkola and Veermans, 2018). The results showed that delayed fading was able to improve the learning outcomes of 6th graders but not 5th graders (among 5th graders, the concrete condition continued to outperform the fading condition).

One design consideration that could be derived from these results is therefore that concrete representations are especially important for younger students and that their benefits may outweigh the potential benefits of generalization through abstract representations even when fading is employed. Generalization benefits might occur only at a later age—there were some indications of this effect in the study on 7th and 8th graders.

6 Virtual Alone or Together with Real?

Traditionally, virtual (computer-based) and real (physical equipment) labs and learning environments have been positioned as competitors, with proponents of real environments emphasizing the importance of tangible experiences for the development of learning and understanding while proponents of virtual environments emphasize that manipulation rather than physicality is at the core of this learning and development and that physicality aside, virtual environments have many advantages over real environments. The main focus of theorizing and empirical research on both sides has been to show that one is better and thus to be preferred. In one of our studies (Jaakkola et al. 2011), we explored the possibilities of combining the two approaches in a 2 × 2 design comparing the two types of instruction mentioned in the previous section (though the previous section discussed only the simulation conditions of this 2011 study) and comparing virtual to a virtual-real combination where students did everything with both the simulation and real equipment (twice the amount of the same work). The instruction dimension extended an earlier study in 2008 that investigated real, virtual and virtual-real environments, but the real condition was excluded due to its showing the least favorable results (Jaakkola and Nurmi 2008).

The results of the 2011 study showed that learning outcomes were better in the combination condition regardless of the instructional support (explicit or implicit), but also that there was an interaction between instruction and learning environments. In the simulation-only condition, explicit instruction prolonged learning time and enhanced learning outcomes (79 min, post = 11.29, pre-post: $p = .013$, d = .78, vs. 66 min, post = 9.75, pre-post: $p = .31$, d = .22), while the same instruction in the combination condition prolonged learning time (similar in magnitude) without enhancing learning outcomes (90 min, post = 12.33, pre-post: $p < .001$, d = 1.24 vs. 73 min, post = 12.67, pre-post: $p < .001$, d = 1.51). In fact, the implicit combination outperformed the explicit simulation condition, and did so within a shorter learning time. This result highlighted once more that it was about the whole learning environment and what it triggered in the learner. Based on explicit or implicit instruction alone with simulation or based on a comparison between explicit simulation and explicit combination, one might have concluded that the students in the lower outcome condition might have learned more if they had spent more time. However, the comparison between implicit combination and explicit simulation showed that it was not that simple.

Implicit instruction alone was not sufficient in the virtual setting to trigger cognitive involvement, and the addition of explicit instruction improved students' performance. Adding real circuits into the combination was even more effective and seemed to make explicit instruction unnecessary. In contrast, adding explicit instruction seemed to slow down the learning process with no apparent learning outcome benefits.

While these results stem from real and virtual environments, it has been argued that the distinction between real and virtual is construed rather than definite (Burbules 2004). More and more of our present day realities are in fact virtual, and trying to maintain a strict dichotomy is therefore no longer helpful. A more flexible framework of thinking that allows virtual experiences to be real experiences allows us to focus more on the nature of the experiences rather than whether they are real or virtual. In the case of the real-virtual setting, experiencing two distinct representations might be of bigger importance for the experience than one being real and the other virtual. If that is the case, the real and virtual experiences from the real-virtual setting described above might very well be incorporated in a similar way in a fully virtual environment, as there is no limitation that prevents combining virtually-real and simulation the same way that real and simulations were combined in this study.

7 How to Combine Multiple Representations?

There are principally two different ways to combine laboratories and simulations (or any mix of multiple representations). In a sequential combination, laboratories and simulations are used in different phases while in a parallel combination, both laboratories and simulations are available at all times. In other words, the main difference is that in a sequential combination, representations are not co-present (most of the time), whereas in a parallel combination, both representations are always co-present. In the studies described here, real and virtual circuits were combined in parallel (Jaakkola and Nurmi 2008; Jaakkola et al. 2010, 2011) while concrete and abstract simulation elements were combined sequentially (bulbs-then-resistors; Jaakkola and Veermans 2015, 2018). While we have not explicitly compared the two types of combination in a single study, our results across studies (see also Gentner et al. 2003) suggest that parallel combination may be the safer option. This may be because parallel combinations are less sensitive to order, while sequential combinations must consider order in the study design (e.g. laboratory or simulation first; concrete or abstract first). Order is far more critical and sensitive in the sequential combination as the representations are used in isolation. Ideally, this entails that each representation should be used when the benefits it can offer learners is higher than the benefits offered by the other representation(s). However, it is not an easy task to determine in practice which representation can offer the highest benefit in a particular context or at a particular moment, because the true benefit is determined by both the informational and computational properties of the representations,[1] which depend on the individual learner's characteristics to some degree.

8 Conclusion

The previous sections presented several key themes from the outcomes of research on a simulation-based learning environment in the domain of electricity. This conclusion will draw on the findings to connect the design of learning environments and the design of educational games. The positive news from this research is that a single learning environment that progresses from more structured in the beginning to more open-ended later on can provide learners across a wide age range with engaging and productive learning experiences. While this matches well with the idea of levels in game design, it is important to focus explicitly on the function of these levels in terms of their learning outcomes (apart from their function for maintaining interest) since the findings across grades from the studies show that there does not seem to be a straightforward relation between interest and learning outcomes. As can be

[1] According to Larkin and Simon (1987), the informational effectiveness of a representation is determined by how much information it contains, whereas the computational effectiveness of a representation is determined by how easily relevant information can be extracted and applied from it.

seen in Fig. 2, 4th graders combined the highest interest with the lowest learning outcomes while in the higher grades where interest was lower, learning outcomes were still considerable. This suggests that the relation between interest and learning outcomes may be a threshold rather than a continuous function in these kinds of learning environments. Game design may not necessarily need to strive to maximize learners' interest, but instead follow a 'good enough' principle. This would leave room for other design decisions that may be more important for obtaining good learning outcomes. One area that may be important to consider in this respect is if there are ways to include elements that trigger reflection, as the studies examined suggest that reflection (either instigated by instruction or by parallel representations) is more important than higher levels of interest. Given the nature of games, instruction might not be the easiest option but parallel (or sequential) representations can be designed into games. This could provide games with elements that trigger reflection without greatly affecting the game's nature.

An area where the design of simulation learning environments could benefit from practices common in game design is in their attention to use and user data for adjusting initial designs. Analyzing use in reference to the intended outcomes makes it more natural to look for aspects in the environment that may be varied and lead to better learner experiences and outcomes. For instance, we found that even a slight delay in the fading from bulbs to resistors led to better learning outcomes and a more efficient learning process with younger learners (5th and 6th graders) in an experimental study. Once an environment has a substantial number of users, the timing of change in sequential designs (e.g. going from concrete to abstract representations) can be randomized in order to find out whether this is an important factor that influences learning outcomes. Adopting this practice in thc building of simulations and serious games could steer designers away from development over a single cycle or a few cycles towards more flexible and continuous iterative development, which may better fit the complicated reality of the interaction between learning environment and learner.

Acknowledgements This work was supported by grants no 116393 and 252580 from the Academy of Finland to the first author and grant no 266189 from the Academy of Finland to the second author.

References

Burbules, N. C. (2004). Rethinking the virtual. *E-Learning, 1*(2), 162–183

European Commission. (2011). *Science education in Europe: National policies, practices and research.*

Fyfe, E. R., McNeil, N. M., Son, J. Y., & Goldstone, R. L. (2014). Concreteness fading in mathematics and science instruction: A systematic review. *Educational Psychology Review, 26*(1), 9–25.

Gentner, D., Loewenstein, J., & Thompson, L. (2003). Learning and transfer: A general role for analogical encoding. *Journal of Educational Psychology, 95*(2), 393–408.

Jaakkola, T. (2012). *Thinking outside the box: Enhancing science teaching by combining (instead of contrasting) laboratory and simulation activities* (Academic Doctoral Dissertation). Annales Universitatis Turkuensis B 352. Turku: Painosalama.

Jaakkola, T., & Nurmi, S. (2008). Fostering elementary school students' understanding of simple electricity by combining simulation and laboratory activities. *Journal of Computer Assisted learning, 24*(4), 271–283.

Jaakkola, T., Nurmi, S., & Lehtinen, E. (2010). Conceptual change in learning electricity: using virtual and concrete external representations simultaneously. In L. Verschaffel, E. De Corte, T. de Jong, & J. Elen (Eds.), *Use of representations in reasoning and problem solving. Analysis and improvement. New Perspectives in Learning and Instruction Series* (pp. 133–152). NY: Routledge.

Jaakkola, T., Nurmi, S., & Veermans, K. (2011). A comparison of students' conceptual understanding of electric circuits in simulation only and simulation-laboratory contexts. *Journal of Research In Science Teaching., 48*(1), 71–93.

Jaakkola, T., & Veermans, K. (2015). Effects of abstract and concrete simulation elements on science learning. *Journal of Computer-Assisted Learning, 31*(4), 300–313.

Jaakkola, T., & Veermans, K. (2018). Exploring the effects of concreteness fading across grades in elementary school science education. *Instructional Science, 46*(2), 185–207.

Larkin, J. H., & Simon, H. A. (1987). Why a diagram is (sometimes) worth ten thousand words. *Cognitive Science, 11*(1), 65–100.

Osborne, J., & Dillon, J. (2008). *Science education in Europe: Critical reflections* (Vol. 13). London: The Nuffield Foundation.

Slavin, R. E., Lake, C., Hanley, P., & Thurston, A. (2014). Experimental evaluations of elementary science programs: A best-evidence synthesis. *Journal of Research in Science Teaching, 51*(7), 870–901. https://doi.org/10.1002/tea.21139.

Tapola, A., Jaakkola, T., & Niemivirta, M. (2014). The influence of achievement goal orientations and task concreteness on situational interest. *The Journal of Experimental Education, 82*(4), 455–479.

Tapola, A., Veermans, M., & Niemivirta, M. (2013). Predictors and outcomes of situational interest during a science learning task. *Instructional Science, 41,* 1047–1064.

Vedder-Weiss, D., & Fortus, D. (2011). Adolescents' declining motivation to learn science: Inevitable or not? *Journal of Research in Science Teaching, 48*(2), 199–216.

Vedder-Weiss, D., & Fortus, D. (2012). Adolescents' declining motivation to learn science: A follow-up study. *Journal of Research in Science Teaching, 49*(9), 1057–1095.

Supporting Conceptual Change in Physics with a Serious Game

Anne van der Linden and Wouter van Joolingen

Abstract Serious games can play a role in physics education, especially in elementary mechanics, as they can provide hands-on experience with force and motion in a simulated environment. In this study, we used a serious three-dimensional immersive game to provide students with an environment in which they needed to search for explanations beyond their preconceptions. We expected that students would see the need for new theories. The goal of the game was for students to direct a ball to a target using forces they could regulate. In a quasi-experimental evaluation between a game group and a traditional group (receiving no game) with 73 participants no significant gain in knowledge was measured in either group. However, students who played the game were more motivated than students who experienced the traditional lesson). Implications for renewed game design and research are discussed.

1 Introduction

1.1 The Role of Preconceptions in Physics Teaching

Students encounter different kinds of motion on a daily basis. In many cases, these can be described on a superficial level using a small set of simple ideas in which terms such as energy are not clearly defined. For more complex kinds of motion, these ideas lead to wrong predictions (Hestenes et al. 1992). For instance, when students are faced with the following problem, they often predict a straight path for the engine: A rocket drifting sideways in outer space is subject to no outside forces; at a certain point, the rocket's engine starts to produce a constant thrust perpendicular to the previous line of movement. However, the correct answer is a parabolic path. One goal of physics

---

A. van der Linden · W. van Joolingen (✉)
Freudenthal Institute, Utrecht University, Utrecht, The Netherlands
e-mail: w.r.vanjoolingen@uu.nl

A. van der Linden
e-mail: a.vanderlinden@uu.nl

Y. Cai et al. (eds.), *VR, Simulations and Serious Games for Education*, Gaming Media and Social Effects, https://doi.org/10.1007/978-981-13-2844-2_3

teaching is to tackle these existing ideas about the laws of motion and promote student comprehension based on physical concepts (Clement 1982). Studies have shown that an effective way to do so in science education is by giving explicit attention to existing preconceptions (Chi et al. 1994; Duit and Treagust 2003; Vosniadou 1994). More specifically, giving preconceptions a central role when teaching the first and second laws of Newton has been shown to yield positive comprehension effects (Muller et al. 2008). Also, using preconceptions as a basis for instruction can contribute to building scientific literacy (De Boer 2000). By creating situations in which preconceptions are no longer sufficient as an explanation, students are exposed to the need to engage in scientific reasoning such as the development of new hypotheses and theories. In the problem-posing approach (Klaassen 1995), this idea is employed: By creating situations in which preconceptions are no longer adequate explanations, learners will see the need for new theories. Students may be able to describe simple motions on a superficial level; however, more complex kinds of motion must be described with more in-depth explanations. Therefore, students need to alter their existing preconceptions and use a formal physical approach to explain more complex kinds of motion. Students will find the need to alter their preconceptions if these are no longer sufficient explanations; therefore, students need to see the behavior of their preconceptions on more complex kinds of motion. However, an important drawback is that teachers have to start from the 'correct' physics situations and problems. In our real world, behavior according to preconceptions cannot be shown, because our world simply behaves according to our known physical laws. If a preconception leads to a certain motion that contradicts the motion according to the physical laws, the motion according to the preconception cannot be shown in reality. To truly show the effect of students' preconceptions, there is need for an environment that is not bound to real physical laws. Such an unreal environment is possible—it is a digital one. Students can be put in an unreal world. There they can set their preconceptions and experience their effects without being bound to the physical limits of this world.

1.2 Serious Games

A serious game is a computer game with the aim of facilitating learning in addition to entertaining users. In a serious game, the entertainment value of video games is used to influence learners' motivation (Charsky 2010). Recent studies show that training with serious games can be more effective in improving knowledge and cognitive skills than training with conventional instructional methods (Sitzmann 2011). The use of a serious game leads to a better retention effect in comparison with conventional instruction methods. Serious games lead to well-structured prior knowledge on which learners can build during their learning careers (Wouters et al. 2013). Wouters et al. (2013) argue that it is possible that immediately after learning from conventional instruction, students are able to remember texts or notes given during instruction, leading to no difference between conventional instruction and game conditions. However, after several days, students benefit more from game conditions due to their deeper level of knowledge processing in games (Kintsch 1998). To make practice with serious games more effective than practice with conventional instruction methods, it is important to supplement the game with other instruction methods such as a class discussion (Wouters et al. 2013). The meta-analysis conducted by Wouters et al. (2013) shows that serious games are more effective in combination with other instruction methods than in isolation. Whilst playing the game, students gain intuitive knowledge; however, in the absence of additional instruction methods, students are not given a chance to verbalize this knowledge and anchor it more profoundly in their knowledge base (Wouters et al. 2008).

1.3 Motivation

A major reason for serious game usage in lessons is that the use of a serious game will likely influence students' motivation. A serious game is intended to be more enjoyable than conventional instruction methods, thus students will likely be more intrinsically motivated to engage in the learning activity (Charsky 2010). Intrinsically motivated behavior refers to doing something because it is inherently interesting or enjoyable (Ryan and Deci 2000). Extrinsic motivation, in contrast with intrinsic motivation, refers to behavior that is driven by external rewards (Brown 2007). Whilst playing a serious game, students can also be extrinsically motivated to learn. For example, students' learning can be rewarded with points in a game, thus they are extrinsically motivated. In addition to the two types of motivation, Ryan and Deci (2000) developed a taxonomy of human motivation where different types of external motivation are defined. A distinction is made between external and internal motivation. External motivation is a controlled form of extrinsic motivation. More autonomous forms of extrinsic motivation are referred to as internal motivation. Since intrinsic motivation is completely autonomous, intrinsic motivation is the ultimate form of internal motivation (Ryan and Deci 2000). An example of internal behavior is when students identify the importance of an activity or when an activity is enjoyable. In this research, the focus lies on internal motivation.

The meta-analysis conducted by Wouters et al. (2013) shows that serious games do not have positive motivational effect on students, contrary to expectations. Wouters et al. (2013) provide several possible reasons for current serious games not being more motivating than conventional instructional methods. The first is that students often lack control over decisions in serious games. As autonomy supports internal motivation (Deci et al. 1991), conditions in the game that limit students' sense of control lead to lower internal motivation. In contrast, when autonomy is stimulated, students are more internally motivated to engage in the learning activity (Connell and Wellborn 1991). The second is that the connection between entertainment design and instructional design is not a natural one. This means that design choices that are good for instructional purposes often have a negative effect on entertainment value (Wouters et al. 2013). For instance, for instructional purposes, it is effective to prompt the student to reflect. The designer could use a pop-up screen to do so. However, this pop-up interrupts the flow of the game which leads to a negative effect on its entertainment value. To counter the disruptive nature of pop-up messages, it is important that the learning goal and the game goal are intertwined. By doing so, students learn without constant reminders of the game's learning purpose. Hence, if the aforementioned ideas are incorporated in a serious game, students will likely be more internally motivated to engage in the learning activity and therefore positive learning outcomes can be expected (Ryan and Deci 2000; Ryan et al. 2006). To measure such internal motivation to engage in an activity, specific statements of the Situational Motivation Scale (SIMS) can be used (Guay et al. 2000). The SIMS measures different kinds of situational motivation, including situational intrinsic motivation and identified regulation which both belong to internal motivation. Situational intrinsic motivation is intrinsic motivation that occurs during the engagement in an activity (Guay et al. 2000). Several existing games were developed with the purpose of improving the comprehension of Newton's laws. Students are often put in an ideal frictionless environment so that their in-game motions represent the effects of forces in a theoretically ideal form. These serious games have shown positive learning outcomes (Koops and Hoevenaar 2011; White 1984). However, these games do not yield a positive motivational effect. Also, due to the ideal theoretical environment of the games, students are not confronted with flaws in their existing ideas, while a confrontation with their preconceptions leads to positive comprehension effects (Chi et al. 1994; Duit and Treagust 2003; Vosniadou 1994).

1.4 Incorporating a Problem-Posing Approach in a Serious Game

By implementing a problem-posing approach in a serious game, students can actually experience the physical effects of their preconceptions. Students are thus directly confronted with their existing ideas of motions, which naturally leads to several problems. Students will find out that some motions are impossible in the world of their own preconceptions. Hence, physical reasoning is stimulated and concept development is needed to adjust students' existing ideas to formal physics.

In a serious game, students are able to experience the effects of their preconceptions. For instance, they can see if their preconceptions lead to an unrealistic movement. Confronted with this unrealistic movement, students encounter the fact that their preconceptions no longer sufficiently explain realistic movements. Therefore, students will find the need to discover new ideas that will lead to explanations for realistic movement. Since this need for explanation will come from the students themselves, students are more likely to engage in the learning activity (Vollebregt et al. 1999).

1.5 The Case Study: Newton's Laws

A very suitable subject for a serious game is Newton's laws. Not only are the laws of motion an important part of the secondary school curriculum, there is also a lot of didactical information available about this subject. The preconceptions of students in the field of mechanics are well known (Driver et al. 1994). There is also a valid research instrument available: the Force Concept Inventory (FCI) (Hestenes et al. 1992). With this information already validated, this research can truly focus on the learning and motivational effects of a developed serious game. Therefore, the aim of this research is to improve students' comprehension and motivation regarding Newton's laws. This leads to the following research question: How can the use of a serious game foster both students' comprehension and their motivation with respect to learning Newton's laws, in comparison with conventional instruction methods?

2 Method

2.1 Research Design

The study used a design-based approach followed by a quasi-experimental evaluation. First design criteria for the serious game were formulated and a first version of the game was developed. The practicality of the game was evaluated by observing several students playing the first version and the game was further developed in a second version. This version was evaluated on the content level of the game and improvements were made to develop the final version. A quasi-experiment using the final version evaluated its effective learning effects and motivational effects.

2.2 Participants

The participants during the design phase included 30 4VWO (Dutch: "*voorbereidend wetenschappelijk onderwijs*"; literally "preparatory scholarly education") students. The participants in the quasi-experimental evaluation included 73 3VWO students between the ages of 14 and 16 (grade 10).

2.3 Instruments

The developed game consists of seven levels. In each level, students need to guide a ball to the finish line (see Fig. 1). They can do this by giving the ball an initial kick—a force (F_{kick}). They also decide if there is another constant force ($F_{constant}$)

Fig. 1 Screenshot of the simulation game with indictions for speed and accellarion. The goal is to reach the goal at the end over increasingly difficult tracks

Table 1 An overview of each level in the game. The first three levels and fifth level are introductory levels

Level	Settings	Friction	Specifications of track
1	F_{kick}: scrollbar	No	Straight and short
2	$F_{constant}$: yes/no	No	Straight and short
3	$F_{constant}$: scrollbar	Yes	Straight and short
4	$F_{constant}$: scrollbar	Yes	Curves, can only finish with realistic physics
5	F_{kick}: scrollbar	No	Introduction of acceleration platforms and deceleration platforms
6	$F_{constant}$: scrollbar	Yes	different roads to finish, curves and platforms, can only finish with realistic physics
7	$F_{constant}$: scrollbar	Yes and no	Different pavements, different roads to finish, curves and platforms, can only finish with realistic physics

working on the ball to keep it moving, which they can set a value for. After the initial kick, students can alter the direction of the ball by giving a small kick to the sides of the ball, perpendicular to the direction of motion. The difficulty of the levels slowly increases. Students start on a straight road with no friction. In later levels, friction is added and curves occur. Platforms on the road are also added where the ball speeds up or slows down. Students lose a level if the ball falls off the road or if the ball becomes stationary. In each level, students are able to collect coins—each worth 10 points—that determine their level score (Table 1).

The first version of the game was evaluated through the observation of students' performances during and between levels in combination with a post-game interview conducted by the researcher. An observational scheme was used in which the researcher noted whether the student finished the level and how and where it went wrong, in the event of failure. The researcher also noted the score per level and whether in-game texts were read. There was also room to note any faults in the level. After each level, several interview questions were asked:

- What do you think about the difficulty of the level?
- Was there anything unclear in the level?
- What do you think about the length of the level?
- Was the control of the ball intuitive?
- Would you play this level again with different settings?

To evaluate the second version of the game, students answered a post-test immediately after playing the game. The post-test included specific items of the FCI (Hestenes et al. 1992). For the quasi-experiment, a pre- and post-test were used which included the same FCI questions. To evaluate the motivational effect of engaging in the learning activity, statements of the SIMS were incorporated in the post-test (Guay et al. 2000).

2.4 *Data Collection and Analysis*

The researcher's observations and participants' answers to interview questions were used to improve the first version of the game. The second version was improved based on the post-test data from the 4VWO students. To evaluate the final version of the game, a quasi-experiment was held in three 3VWO classes. These classes divided naturally into three groups. The first group (the control) experienced a traditional lesson: They listened to a classroom instruction then completed assignments and revised those assignments based on feedback by the teacher. The second group played the game without other classroom activities. The third group started the lesson by playing the game, followed by a classroom discussion. This discussion included images from several levels of the game. All groups started with a pre-test and ended the experiment with a post-test. The duration of the experiment in all groups was a single lesson of 40 min including the pre- and post-test. All experiments were held on the same day.

One week after the experiment, the control group also played the game. Instead of a post-game classroom discussion, the students received a worksheet with questions they needed to fill in after each level. The motivational effect of playing the game whilst answering a worksheet was thus evaluated.

Table 2 The results of the evaluation of the first version of the game and the improvements that were made for the second version

Observation/participant Response	Improvement for the 2nd version
When the ball falls off the track, it keeps moving	When the ball falls off the track, the ball comes to a stop and students are able to restart the level
Levels 4 and 6 are too difficult	The initial force (F_{kick}) in level 4 and 6 was lowered
The scrollbar does not work properly for the setting of forces	Scrollbar was fixed
In-game text is mostly not read	In-game text was shortened
If in-game text was not read, setting the forces is unclear	In-game text and setting were featured in the same pop-up
The deceleration platforms do not work	Deceleration platforms were fixed

3 Results

There were several observations and participant responses during the evaluation of the first version of the game that led to game improvements (Table 2).

The results from the evaluation of the second version of the game led to improvements to the final game. In the second version, a short kick animation was shown when the player kicked the ball, for instance at the start of a level. On the post-test for the second version, 92.3% of students correctly answered a question about the meaning of this animation. The group scored 52% on the FCI questions of the post-test. However, the students scored the lowest on the questions about direction of a moving object. To provide more clarification in the final game version for the influence of a sideways kick on the ball motion, a kick animation was added to the sides of the ball when the direction of the ball was changed via the arrows on the keyboard. With these added animations, students were more likely to see changes in direction due to a kick, instead of an internal steering system. To give students more insight on the effects of a $F_{constant}$ in comparison with no $F_{constant}$, it was decided that level 2 had to be played twice: once with an $F_{constant}$ working on the ball and another time without $F_{constant}$. This way, students could see the effects of such a force at least once and use this information in the later levels where a realistic movement had to be made.

Most students were not able to finish all levels. In the second version's last level (level 8), students were able to set F_{kick}, $F_{constant}$, and the mass of the ball. This level was deleted in the final version of the game (Table 3).

A paired samples t-test was performed to examine the mean differences between the pre-test and the post-test of each group. For the control group, there was no significant difference in the pre-test ($M = 2.40$, $SD = 1.14$) and post-test ($M = 2.88$, $SD = 1.301$) scores; $t(25) = 0.755$, $p = .252$. For the game group, there was also no significant difference in the pre-test ($M = 2.19$, $SD = 0.895$) and post-test ($M = 2.00$, $SD = 1.020$) scores; $t(24) = -1.174$, $p = .446$. Finally, no significant differences

Table 3 The results of the pre- and post-test (with a minimal value of 0 and a maximal value of 6) and the results of the motivation scale (with a minimal value of −2 and a maximal value of 2)

	Control group Mean(SD)	Game group Mean(SD)	Experimental group Mean(SD)
Pre-test	2.40 (1.258)	2.19 (.895)	2.32 (1.041)
Post-test	2.88 (1.301)[a]	2.00 (1.020)[b]	2.68 (1.171)[a]
Motivation	−.3125 (.933)[a]	.7019 (1.061)[b]	.9773 (.873)[b]

Statistically significant differences are indicated with an index. If the indices of two means in a row are different (i.e. a and b), the two means significantly differ from each other. If the indices of two means in a row are identical, there is no statistically significant difference between the means

were found between the pre-test (M = 2.32, SD = 1.041) and post-test (M = 2.68, SD = 1.171) scores of the experimental group; t(21) = −1.093, $p = .287$. Overall, the intervention had no significant effect on the learning results in all three groups.

A post hoc analysis (Fisher's Least Significant Difference [LSD]) was performed to examine the mean difference between the post-tests. The results show that a significant difference can be found at the .05 level between the control group and the game group ($p = .009$). Furthermore, the Cohen's effect size value ($d = -.753$) is. A significant difference can also be found between the game group and the experimental group ($p = .048$) with a moderate to high effect size ($d = .619$). However, no statistical significant differences were found between the control group and the experimental group ($p = .563$). These results show the importance of embedding the game in a lesson. The final results of the students who only played the game are significantly lower than those of students who experienced a traditional lesson or students who discussed the game afterwards.

A second post hoc analysis (LSD) was performed to examine motivational effects. The results show that a significant difference can be found at the 0.05 level between the control group and the game group ($p < .001$). The effect size is high ($d = 1.02$). A significant difference can also be found between the control group and the experimental group ($p < .001$, $d = 1.43$). In contrast, no statistically significant difference can be found between the game group and the experimental group ($p = .327$). These results support the motivational effect of the game: Both groups who played the game showed a significant motivational effect in comparison with the group who experienced a traditional lesson. When students filled out a worksheet whilst playing the game, a significant motivational effect can be found in comparison with a traditional lesson ($p < .001$). Further, the Cohen's effect size value ($d = 2.47$) suggested a very high effect size. The use of a worksheet also produced a statistically significant difference in comparison with the game group ($p = .012$) with a high effect size ($d = .911$). However, no significant differences can be found in motivation between using the worksheet and a class discussion following the game ($p = .131$).

4 Conclusions and Discussion

The aim of this study was to improve both students' comprehension and their motivation regarding Newton's laws. To achieve the research goal, the following research question was answered: How can the use of a serious game foster students' comprehension and motivation with respect to learning Newton's laws in comparison with conventional instruction methods?

With the current version of the game, participants' comprehension of Newton's laws does not improve more than from a traditional lesson. There was no significant difference found between the pre- and post-tests of all three conditions. This means that in all groups, the learning effects—if any—were low. This could possibly be due to the very short intervention time of 40 min. As both pre- and post-tests were conducted during that time period, the effective intervention time was only 25 min. Results show that the traditional lesson is about as effective as the experimental lesson regarding learning effect. That learning effects do not differ between the students who played the game and the students who practiced with conventional instruction methods corresponds with the results of Wouters et al. (Wouters et al. 2013). Students completed the post-test directly after the intervention, so measuring retention effects was not possible. In addition, it is important to embed the game in a lesson to improve comprehension. Students who only played the game scored significantly lower on the post-test than both other groups, which again corresponds with the results of Wouters et al. (2013).

Students who played the game as a lesson activity were clearly more motivated than students who received traditional instruction methods. To achieve this motivational effect, several criteria were implemented in the game. Students were able to incorporate their own ideas about motion and forces in the game, and they could instantly see the effects of those ideas and come to a conclusion about how realistic their ideas were. Then they could alter their ideas and try to achieve a realistic movement. The learning goal and the game goal are intertwined with each other. Lastly, students were able to make their own decisions in the game. They could set their own rules for motion and there were even multiple routes to the finish line in some levels. To foster a motivational effect, it was expected that not disturbing the flow of the game would be important. However, even when using a worksheet while playing the game, a significant motivational effect was found in comparison with a traditional lesson.

Before using the developed game in further research, it should be noted that the game itself needs some improvements. With the current version of the game, it was possible for students who had some experience of the game to complete it with the use of nonrealistic physics. This outcome should be made impossible. Also, before each level in the game, short texts appeared with information on how to play the game. However, students generally did not read those texts while playing the game, so it took them some time to figure out what they were supposed to do. Lastly, students scored lowest on the questions about direction of motion. Replacing the currently narrow track with a broader track would allow students to more clearly see the influence of a kick on the ball's movement.

To achieve a comprehension effect regarding Newton's laws, several aspects need to be taken into account for further research. It has been shown that just playing the game is not an effective learning method. Therefore, the game should be embedded in a series of lessons, lengthening the intervention time and thus solving the earlier stated problem of the short intervention time as well. To gain more insight into students' reasoning and comprehension of the subject, their reasoning should be made explicit during or after playing the game. To measure a learning effect, a retention measurement should be performed. Wouters et al. (2013) argue that it is possible that immediately after learning from conventional instruction, students are able to remember texts or notes given during instruction, leading to no difference between conventional instruction and game conditions. However, after several days students benefit more from game conditions, due to the fact that in a game students process a deeper level of knowledge (Kintsch 1998). To improve students' comprehension and to achieve a learning effect, students need some guidance whilst playing the game, since they generally did not read in-game texts. A worksheet is one possibility, but are there more effective methods? What should the role of the teacher be in the lessons? To answer these questions, further research is needed.

References

Brown, L. V. (2007). *Psychology of motivation*. New York: Nova Science Publishers Inc.

Charsky, D. (2010). From edutainment to serious games: A change in the use of game characteristics. *Games and Culture, 000*(00), 1–22.

Chi, M., Slotta, J., & De Leeuw, N. (1994). From things to processes: A theory of conceptual change for learning. *Learning and Instruction, 4,* 27–43.

Clement, J. (1982). Students' preconceptions in introductory mechanics. *American Journal of Physics, 50*(1), 66–71.

Connell, J. P., & Wellborn, J. G. (1991). *Competence, autonomy, and relatedness: A motivational analysis of self-system processes*. Hillsdale: Lawrence Erlbaum Associates.

De Boer, G. E. (2000). Scientific literacy: Another look at its historical and contemporary meanings and its relationship to science education reform. *Journal of Research in Science Teaching, 37*(6), 582–601.

Deci, E., Vallerand, R., Pelletier, L., & Ryan, R. (1991). Motivation and education: The Self-determination perspective. *Educational Psychologist, 26*(3&4), 325–346.

Driver, R., Squires, A., Rushworth, P., & Wood-Robinson, V. (1994). *Making sense of secondary Science: Research into children's ideas*. Oxon: Routledge.

Duit, R., & Treagust, D. (2003). Conceptual change: A powerful framework for improving science teaching and learning. *International Journal of Science Education, 25,* 671–688.

Guay, F., Vallerand, R. J., & Blanchard, C. (2000). On the assessment of situational intrinsic and extrinsic motivation: The situational motivation scale (SIMS). *Motivation and Emotion, 24*(3), 175–213.

Hestenes, D., Wells, M., & Swackhamer, G. (1992). Force concept inventory. *The Physics Teacher, 30,* 141–158.

Kintsch, W. (1998). *Comprehension: A paradigm for cognition*. New York: NY: Cambridge University Press.

Klaassen, C. (1995). *Klaassen, C.W.J.M. (1995), A problem-posing approach to teaching the topic of radiation.* Utrecht: Cdβ Press.

Koops, M., & Hoevenaar, M. (2011). Educatieve game gaat misconcept te lijf. *Nvox, 8,* 372–373.

Muller, D., Bewes, J., Sharma, M., & Reimann, P. (2008). Saying the wrong thing: Improving learning with multimedia by including misconceptions. *Journal of Computer Assisted learning, 24,* 144–155.

Ryan, R. M., Rigby, C. S., & Przybylski, A. (2006). The motivational pull of video games: A self-determination theory approach. *Motivation and Emotion, 30,* 347–365.

Ryan, R., & Deci, E. (2000). Intrinsic and extrinsic motivations: Classic definitions and new directions. *Contemporary Educational Psychology, 25,* 54–68.

Sitzmann, T. (2011). meta-analytic examination of the instructional. *Personnel Psychology, 64,* 489–528.

Vollebregt, M., Klaassen, K., Genseberger, R., & Lijnse, P. (1999). *Leerlingen motiveren via probleemstellend onderwijs. Nvox, 7,* 339–341.

Vosniadou, S. (1994). Capturing and modeling the process of conceptual change. *Learning and Instruction, 4,* 45–69.

White, B. (1984). Designing computer games to help physics students understand Newton's laws of motion. *Cognition and Instruction, 1,* 69–108.

Wouters, P., Paas, F., & van Merriënboer, J. J. (2008). How to optimize learning from animated models: A review of guidelines base on cognitive load. *Review of Educational Research, 78,* 645–675.

Wouters, P., van Nimwegen, C., van Oostendorp H., & Spek, E. D. (2013). A meta-analysis of the cognitive and motivational effects of serious games. *Journal of Educational Psychology* (Advance online publication).

Evaluation of a Re-designed Framework for Embodied Cognition Math Games

Jonathan D. L. Casano, Hannah Tee, Jenilyn L. Agapito, Ivon Arroyo and Ma. Mercedes T. Rodrigo

Abstract Embodied cognition posits that the development of thinking skills is distributed among mind, senses, and the environment. Research in this field has resulted into the development of applications in different areas including mathematics. This paper reports one part of a larger series of studies on the design and implementation of embodied cognition-based mathematics educational systems. We describe the evaluation of a game called *Estimate It!*, a wearable-based game for teaching measurement estimation and geometry. Experts were invited to evaluate the game, resulting in a generally positive rating. The game's collaborative nature, its hands-on way of teaching estimation, and the incorporation of technology were seen as promising points. Infrastructure readiness, classroom control, and adjustment to the new technology were areas of concern.

1 Introduction

Cognition has traditionally been viewed through a narrow perspective in which the body has mostly sensory and motor functions, subordinate to central cognitive processing. In recent years though, the idea of cognition has broadened: It is now believed to be distributed among mind, senses, motor capabilities, and social interactions

J. D. L. Casano (✉) · H. Tee · J. L. Agapito · Ma. M. T. Rodrigo
Ateneo de Manila University, Katipunan Ave, Quezon City, Metro Manila 1108, Philippines
e-mail: jonathancasano@gmail.com; jcasano@gbox.adnu.edu.ph

H. Tee
e-mail: hannah.tee@obf.ateneo.edu

J. L. Agapito
e-mail: jen.agapito@gmail.com

Ma. M. T. Rodrigo
e-mail: mrodrigo@ateneo.edu

I. Arroyo
Worcester Polytechnic Institute, 100 Institute Rd, Worcester, MA 01609, USA
e-mail: iarroyo@wpi.edu

Y. Cai et al. (eds.), *VR, Simulations and Serious Games for Education*, Gaming Media and Social Effects, https://doi.org/10.1007/978-981-13-2844-2_4

(Wilson and Foglia 2011). This theory of cognition, called embodied cognition, assumes that sensory perceptions, motor functions, and sociocultural contexts shape the structure and development of thinking skills including mathematical abilities and higher-ordered abstract reasoning, as well as sense-making in general (Hornecker and Buur 2001; Redish and Kuo 2015).

Research into embodied cognition has resulted in the development of applications in mathematics (Link et al. 2013), ecology and environment (Esteves 2012), problem solving (Malinverni and Burguès 2015), and other disciplines. Physically interacting with applications can involve physical tools that allow the manipulation of information and the incorporation of speech and gesture as inputs, among others. For example, Clavier is a walkable keyboard and audio device stalled along a foot path (Hornecker and Buur 2001). As pedestrians walk along the path, they trigger colorful lights and drum beats.

Some of these studies have affirmed embodied cognition's pedagogical value, finding that children gain a better grasp of abstract representations such as number lines when asked to walk to a position instead of pointing to a value on a sheet of paper or on a blackboard (Link et al. 2013), for example. In the context of collaborative learning, embodied cognition has socio-affective impacts, creating increased perception of cooperation and reciprocal learning (Malinverni and Burguès 2015).

Some researchers remain circumspect though. Despite the positive findings from the studies mentioned in the previous paragraphs, there is still uncertainty regarding what circumstances embodied cognition yields more intellectual or social profit than other pedagogical methods under (Esteves et al. 2013). Furthermore, the processes of designing and evaluating embodied cognition-based educational applications still lack guidelines (Esteves 2012, Schaper et al. 2014). There is a dearth of literature providing designers with frameworks for building real-world interactions (Hornecker and Buur 2001).

In this paper, we report part of a larger study about the design and implementation of embodied cognition of mathematics educational systems. We describe *Estimate it!*, a wearable-based game for teaching size estimation. We then summarize the results of an expert review of the application. We attempt to answer the following questions:

(1) In what ways is this method of teaching estimation superior (or inferior) to traditional ones?
(2) What student characteristics (intellectual, social, and cultural) might make the application suitable (or not suitable) for teaching estimation?

2 Prior Work

The work described in this paper is the continuation of work described in Arroyo 2014. The Cyberlearning Watch is a device that students wear on their wrist to receive clues that help them search for geometric pieces that may be hidden. As they complete the tasks, they receive accuracy feedback via the watch's display, buzzers, or lights, making the student's experience immersive and interesting.

The watches were originally made from Arduino Uno Microcontrollers (SparkFun) connected to a WiFly wireless module that was in communication with a central server. The server, located in situ, selected questions for each student, kept score, and logged events for teachers and students to analyze later.

The main limitation of this implementation was that each watch had to be custom-built. Once the proof of concept was tested, the next step was to improve the devices by adding NFC scanners and considering commercially available cellular phones and wearables.

This chapter describes the evaluation of the original Arduino-based watch migrated to Android-based and Tizen-based system. To demonstrate the system's functionality, we specifically migrated the game *Estimate It!* (Rountree 2015) and asked a group of grade school mathematics teachers to evaluate it. In *Estimate It!*, students receive clues such as *Find a rectangle 2′ wide by 14′ long* or *Find a grid with rectangles approximately 3.25 in. by 2.25 in.* and have to traverse a physical space searching for objects, being provided with specific measurement tools such as a 12-in. dowel without markings (which encourages rounding when participants partition it or use it to measure larger objects). The participants are given 20 min to find as many objects as possible using the clues given, and can work in teams with watches semi-synchronized to each other (Fig. 1).

3 Expert Evaluation

The goal of the expert evaluation was to obtain feedback from grade school mathematics teachers regarding the feasibility of use and implementation of *Estimate It!* in classrooms with real students. More specifically, we aimed to answer the research questions posed at the end of the first section.

3.1 Participants

A total of fourteen (14) Mathematics teachers from Ateneo de Manila University participated in the evaluation: eight females and six males. Eleven (11) of them handled grade school classes, one handled secondary school classes, while the other two were college instructors. They were of varying ages ranging from 23 to 64 years. Their teaching experiences were also diverse, with the youngest participant having taught for a year while the most experienced teacher had 27 years of experience.

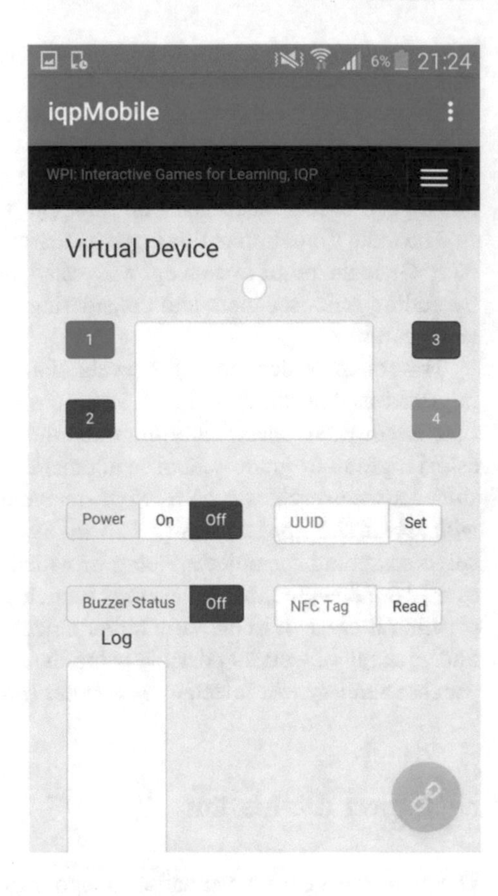

Fig. 1 A screenshot of the "Estimate It!" game running on a Samsung Galaxy S5 (Android Lollipop)

3.2 Methods

The evaluation took place in a classroom equipped with wireless network connection. The server, which consisted of a laptop running on Linux that ran the web server software, was set up in a designated area in the venue. Objects that had been tagged with both NFC tags and sequences of color codes (Fig. 2) were scattered around the classroom. The geometric pieces were everyday objects (e.g. a book—rectangular prism, a ball—sphere) that depicted the sought for objects, described by the clues given by the application. Two Samsung Galaxy S5 smartphones and a Samsung Gear S2 SmartWatch were used during the evaluation. These devices were pre-installed with the game and likewise made to connect to the wireless network.

The experiment team consisted of two people who played the roles of facilitator and test monitor. Upon welcoming the participants, the facilitator described the game, how it worked, and how it might be implemented in their classes. The goals of the evaluation and the method by which it would be conducted were explained. The teachers were asked to fill out a demographics questionnaire before beginning their gameplay.

Fig. 2 Geometric objects used in the evaluation of 'Estimate It' (left). Color codes and NFC tags attached to geometric objects (right)

Estimate It! was designed to be played with nine participants split into three teams. During the evaluation, the teachers acted as students of a team. As the game commenced, the participants awaited clues in their devices and moved around the room to find the objects described. The facilitator and the test monitor followed them around throughout the game, explaining their progress and answering questions as necessary. Once the game was completed, the teachers were asked to answer a debriefing questionnaire.

3.3 The Debriefing Questionnaire

The debriefing questionnaire used a five-point scale, with possible responses going from *Strongly Disagree* (1) to *Strongly Agree* (5). Questions were derived from the criteria described by Whitton 2009 for effective educational design of game-based learning applications. Items relevant to the purpose of the experiment were re-constructed into questions. Table 1 shows the items classified into four groups.

Follow-up questions asking about what aspects the evaluators liked the most and least about the game as well as their insights on its advantages and disadvantages over current teaching methods were also included.

3.4 Findings

In general, most of the experts gave highly positive evaluations of the game in the questionnaire. The aggregated frequencies of each question have been summarized in Fig. 3. The dispersion of responses shows typically average to little variance except for Question 5 (Q5) in the *Engenders Engagement* group of questions (see Table 1),

Table 1 Effectiveness of educational design of game-based learning applications questionnaire, Whitton (2009)

Active learning support	
1	The game encourages exploration, problem-solving, and inquiry
2	The game provides opportunities for collaboration
3	The game provides opportunities to test ideas and gain feedback
4	The game provides opportunities for practice and consolidation of knowledge
5	Game goals are aligned with the subject's learning goals
Engenders engagement	
1	Goals of the game are clear and achievable
2	The game supports a high level of interactivity
3	The game stimulates curiosity and puzzlement
4	The game establishes the application of estimation in the real world
5	Game levels are appropriate and challenging
Appropriateness	
1	Game goals are aligned with the curriculum
2	Game goals are aligned with the subject's learning outcomes and assessment
3	A game-based approach is applicable for teaching mathematics, specifically estimation
4	The game could be played within the allotted class period
Classroom use	
1	I think the game will help students learn estimation
2	I think my students will find this game fun
3	I think I would use this game in my classes
4	I would recommend this game to my colleagues

in which one of the evaluators strongly disagreed with the statement about the game levels being appropriate and challenging. According to her, the game design did not reflect cognitive levels that indicate mathematical understanding, particularly estimation. She suggested that the levels be created based on the different types of estimation (e.g. establishing a difference between eyeball estimation and computational estimation), since one type is higher than the other according to Bloom's Taxonomy. However, evaluations were generally positive, including the answers given in the follow-up questions which are summarized next.

The results shown in Fig. 3 also reveal significant circumspection and noncommittance on the part of teachers in Question 3 (Q3) of the *Classroom Use* group of questions, which is about whether teachers would be likely to use the game in their classes. After digging into their comments, we found that the reason may be the perceived difficulty in organizing the game for a big class, as well as the extra time needed to set up the game. One participant specifically raised a concern about grade-school boys needing lengthier orientations to become acquainted with the activity and its interface. Thus, future implementations of wearable embodied games such

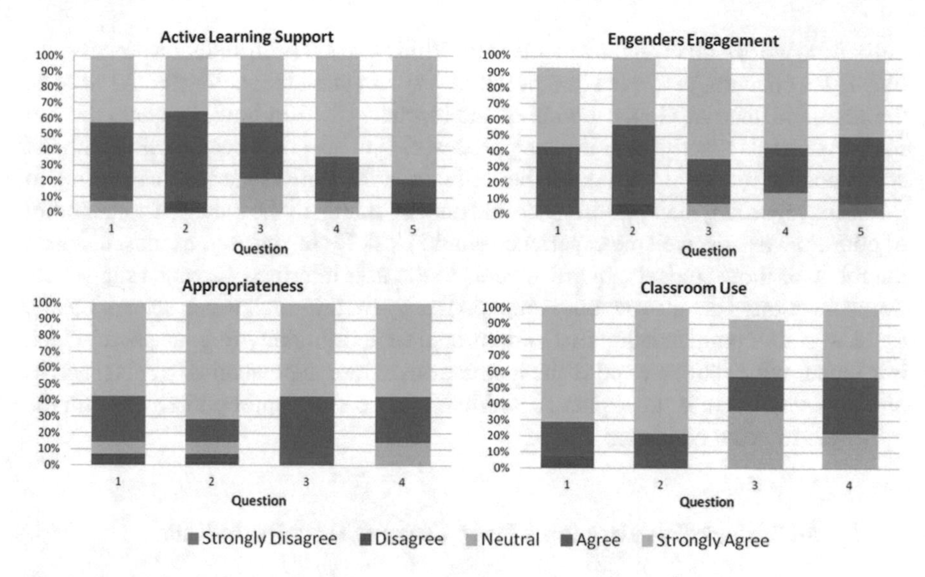

Fig. 3 A graph presenting the aggregate percentages of responses from the evaluators

as *Estimate it!* must consider the implementation aspect as well as the mechanisms that make it simpler for teachers to implement in their classrooms.

Participant teachers found the goals of the game clear and achievable (Q1:Engenders Engagement). All expressed their agreement that a game-based approach was applicable for teaching estimation (Q3:Appropriateness). Pointing out that the generation at present was very much into interactive technologies, the teachers concurred that kids would find the game fun (Q2:Classroom Use). They saw that the use of gadgets and games would entice the students more because they were growing up in the age of technology and were more visually oriented.

3.4.1 Most/Least Likeable Aspects of the Game

The evaluators liked the idea of allowing students to collaborate with one another and the opportunity provided for interaction not just with classmates but with the environment as well. The method provides a more concrete and pictorial way of presenting the concept of estimation. The use of common, ordinary household or office items was seen as a strong point because of their accessibility. Furthermore, this interaction with familiar surrounding objects was judged as being promising in its potential to allow students to physically see and associate measurements and lengths in the things that they saw daily.

On the other hand, the game's high dependence on technology was concerning for teachers. The need for a stable Internet connection and the presumption that students should already be familiar with smartphones and smartwatches led one of the evaluators to speculate that the game might be useful for private schools but was unlikely to work well in public schools in the Philippines. The installation process was likewise a concern, perceived as a possible way in which technology could impede the lesson instead of aiding it. One of the teachers disliked how the game did not require students to write their thinking processes (i.e. how they decided which object corresponded to the instruction). In her opinion, writing gives students something to reflect on regarding their performance and understanding of the matter. Another point of concern was how well the experience would work for big classes, as most teachers taught 40 or more students. Improvements to the user interface were also suggested. Some mentioned the instructions being too wordy for the small watch screen. Lastly, there was a recommendation that certain unclear terminology (e.g. a sphere which is x inches wide) be reviewed in the instructions. More mathematical terms (such as *circumference* to refer to a sphere's *width*) could be more appropriately used in the messages to avoid confusion.

3.4.2 Advantages/Disadvantages Over Current Teaching Methods

According to the evaluators, the use of technology would be attractive to students. Hence, they assumed that the technology and the *Estimate It!* game would be a motivator in the classroom. Some reported that the interactive nature of *Estimate It!* made it more fun and exciting than the usual classroom lecture. The game's stimulation of collaboration by allowing students to work within a team to achieve a competitive goal was a positive. Another positive aspect highlighted was that it encourages students to work and think fast within a given time frame. It was mentioned that it would help the audience see and appreciate math. The dynamic nature of the game was highlighted: The game allows students to *get their hands dirty* as opposed to passively sitting down during discussions.

The logistics and technological requirements, however, were gray areas. Respondents felt that connectivity, being a key factor in such a technological intervention, might hinder implementation since schools might not be ready for it infrastructure-wise. Further, the need to provide smartphones and/or smartwatches could be an issue affecting feasibility. It was mentioned that setting up may be tedious to some teachers and might take too much time. Also, some teachers were concerned about the need for greater classroom management than in the usual classroom setting. Lastly, there was concern that the gadgets might divert the attention of the student from learning the concept to playing with the device (Fig. 4).

3.5 Difficulties Encountered

Initially, we planned to conduct tests using the version of the game uploaded to the Heroku server. However, we discovered that the school network only had Port 80 open and did not allow the establishment of communication through other ports. As a result, we had to reschedule one of the test sessions and resorted to testing

Fig. 4 Math teacher being assisted by the facilitator in scanning Geometric objects around the classroom using a Samsung Galaxy S5 phone during the evaluation

via Localhost (a server running locally in a laptop or computer in the classroom, serving WiFly). Also, since the game is highly dependent on an Internet connection, the game response was affected whenever the Wi-Fi connection in the venue was unstable. During one of the tests, we had problems with the connection and had to invite the participant to step outside so that we could get a good network signal.

4 Student Evaluation

In addition to the expert evaluation, a student evaluation was also conducted to determine the usability of the smartwatch application *Estimate It!* from the perspective of student users. The usability was measured by three main factors: (1) the extent to which students learned by using the app, (2) the ease with which the game can be used in a classroom environment, and (3) the problems and issues with the game's interface design.

4.1 Participants

Seven students—consisting of four boys and three girls, ranging from 4th to 6th grade—participated in this study. All students were familiar with the use of touchscreen-based technology. The students were also fluent in English, minimizing misunderstandings caused by language barriers.

4.2 Method

Each student placed in a controlled environment with a set number of people. A facilitator briefed the student on the app they were testing and what to do during the study. Each student took a pre-test before using the app in order to determine the students' prior levels of proficiency in the domain of estimation. The test was paper-based, consisting of two multiple-choice items, a selection item, and one or two word problems. The content of the test was modelled after tests made for a previous usability study for the application.

The student then played the app on the smartwatch until they finished the game or chose to give up. Each question in the game constituted one task in the study. In order to complete the game, the student had to answer all five questions correctly. As three users are required for progress through the levels, an observer played as the student's other teammates. This portion of the study was filmed so that it may be reviewed at a later time.

During this time, a think-aloud protocol was practiced. Additionally, observers noted down the following data to investigate the last two goals of the study. For the goal of determining the app's feasibility within a classroom environment, two metrics were noted: whether the student was able to complete each task (task completion), and the time it took them to do so (time-on-task). The time per task measurements were added to get the total time it took to complete all tasks. For the app to be deemed feasible for class use, all students had to finish all tasks, and the resulting time could not exceed 30 min for any student.

For the goal of determining the ease of interaction, another four metrics were observed. The first was task completion. Similar to the previous study goal, all students had to complete all tasks for the app to be deemed easy to use.

Another metric observed was the errors made, testing whether the interface was easy to navigate. During a previous usability study on the app from the perspective of the instructor, participants noted that the application's interface was difficult to navigate, with buttons on the smartwatch being too small to push. This had led to accidental button pushes, producing wrong answers. Similar error instances were expected to mostly occur within the first half of playing the game.

Student behavior was also noted. Behavioral instances of students were observed in 15-s intervals, with the dominant emotion among engagement, confusion, boredom, and frustration being noted for each interval. The total instances per student

Table 2 Pre- and post-test results

Participant	Pre-test (%)	Post-test (%)
1	63.64	50.00
2	90.91	70.00
3	100.00	50.00
4	63.64	50.00
5	100.00	100.00
6	81.82	60.00
7	81.82	60.00

were then subject to a one-way ANOVA. Lastly, any issues which the student encountered during play were noted.

The facilitator could step into assist the student whenever one of the following occurred: (1) The student asked for help; (2) the student had gotten the question wrong thrice; and (3) the student had taken more than five minutes on a question.

After the activity, the student took another written test. Questions in this post-test were similar but not identical to those in the pre-test (i.e. did not have the same numbers). The results of the pre- and post-tests were subjected to a paired T-test (see Table 2 for a summarized list of the metrics) to determine if there was a significant difference correlating to the playing of the game, thus revealing if learning had occurred.

After the post-test, students were interviewed and asked a number of questions to get feedback on the app and the task they needed to complete (self-report).

4.3 Results

Although all students were affected by a faulty network connection which caused delayed responsiveness, Students 1 and 2 were the most affected, leading to them not finishing all five questions of the game. Out of all the students, four opted to stop using the smartwatch and continue playing the game on a smartphone instead.

During student playthroughs, the database had to be reset multiple times due to unresponsiveness. This occurred mostly during the progression to level 2. There were instances wherein, although all three users in the team successfully input the correct answer, the app did not allow them to move to the next level, instead getting stuck on the message: *Now take your shape to the starting area and wait for the rest of your team.* From this point onwards, the game became unplayable. Resetting the database fixed this, but returned the student to the beginning of the game. This led to frustration on the part of the student as some students had to keep answering the same question, only to get stuck immediately after.

Table 3 Task completion and time on task

	Task					
Participant	1	2	3	4	5	Total Time
1	7:57:00	3:26:00	DNF	DNF	DNF	11:23:00
2	6:17:00	6:27:00	0:39:00	0:55:00	DNF	14:18:00
3	8:16:00	6:47:00	3:52:00	3:25:00	1:01:00	23:21:00
4	14:01:00	2:03:00	2:42:00	4:50:00	4:21:00	27:57:00
5	5:20:00	7:06:00	1:19:00	2:04:00	1:54:00	17:43:00
6 & 7	0:50:00	1:22:00	0:35:00	1:05:00	0:55:00	4:47:00

4.3.1 Pre-test and Post-test Results

As seen in Table 2, all students did worse in the post-test than they did in the pre-test. The T-test showed that there were significant differences in the results of the pre-test from the post-test, implying that not only was there minimal learning involved, but that the results of the post-test were significantly lower than the pre-test. This could have been because the game did not provide opportunities for students to improve on their understanding of word problems (the part of the test which most students were unable to answer correctly). It could also indicate the students' loss of interest on the study.

4.3.2 Completion and Time on Task

As seen in Table 3, Students 1 and 2 were unable to complete the game, with Student 1 only reaching the second question and Student 2 reaching the fourth. This was due mostly to the connectivity issues of the application, as the application would sometimes stop responding and the team data would not synchronize at times. After multiple attempts at getting a network connection stable enough to continue the game, the two students gave up attempting to finish answering all the questions. This implies that the game in its current condition cannot be reliably played within a regular classroom setting, and that it must be improved to enable students to use it.

Moreover, while the final five students were able to complete the game within the 30-min target, it should be noted that their recorded times do not include the time spent on getting the application to respond. The total length of each playthrough (including the downtime caused by connectivity issues) ranged from 40 min to an hour. As such, a full game may not be finished within the 30-min limit given the current build of the application.

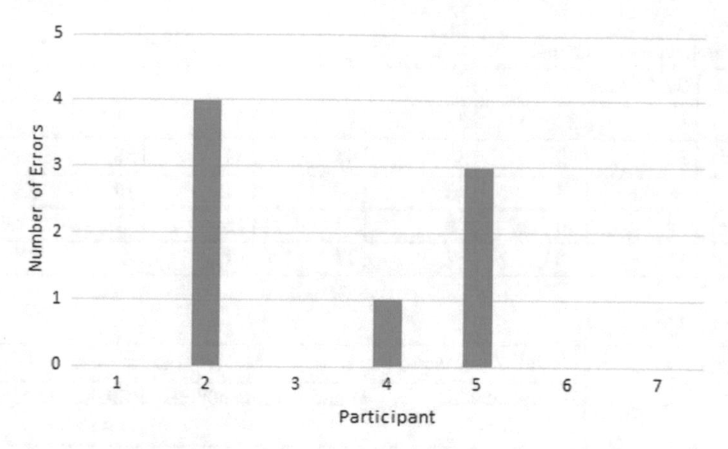

Fig. 5 Number of errors

4.3.3 Errors

For this metric, the only instances of error recorded were when the student verbally indicated that they meant to press another button.

As opposed to the anticipated result from the previous usability test, few errors occurred while the students played, with some students expressing no errors at all. This means that button mispresses are not as grave of an issue as originally thought. Moreover, all errors occurred within the first half of the playing time, implying that any errors caused by button presses can be overcome as time passes. Despite this, some students did complain about small buttons and suggested having larger ones during the self-report (Fig. 5).

4.3.4 Issues

See Table 5.

4.3.5 Behavior Indicators

Unlike the data for Task Completion and Time-on-Task, data on student behavior included the downtimes caused by app unresponsiveness and unstable network connection, to take into consideration students' reaction to unexpected events. Additionally, while the students played for 40 min to an hour, student behavior was only observed during the first 15 min of each playthrough.

From Table 4, it can be seen that the most commonly occurring behavior is the feeling of being engaged within the game. A one-way ANOVA shows that there is a significant difference among the behaviors ($p < 0.0001$). A further Tukey mean

Table 4 Behavior of Students

Behavior	Participants						Average
	1	2	3	4	5	6&7	
Engaged	37	43	38	45	34	38	39.1666667
Bored	8	3	6	0	1	13	5.1666667
Frustrated	4	6	9	2	18	7	7.6666667
Confused	11	8	7	13	7	2	8

Table 5 Usability issues

Severity	Usability issues		
	Faulty network connection	Hard to press buttons	Takes too long to load; app stalls
Low	1	0	2
Medium	2	1	2
High	5	2	3

separation test indicates that it is only the engaged behavior which is statistically significant. This implies that students were able to engage with the application despite the issues they encountered (Table 5).

Most issues occurred with the faulty network connection. Measures were taken at the beginning of the study to reduce network connection issues by having three different networks to connect to, but all three networks still caused the application to load slowly and become unresponsive. This led to long playing times and the inability to complete tasks for some students.

Some students also had difficulty pressing the buttons. Specifically, the buttons were so small that the students could not see the icons while their fingers hovered over the buttons. Consequently, students were unsure whether the button was actually being pressed.

Finally, the long loading times and random stalling of the application might have been due to the network connectivity. The long loading time caused students to press buttons multiple times under the assumption that the watch had not detected their finger when they pressed the button. This led to double presses which made the students' answers wrong.

4.3.6 Self-report

The ratings of the students were generally positive. They found the application fun and wanted to use it in the classroom with their classmates. The only dissenters were the final two participants, who stated that *[their] class was noisy enough without it.* This is an important point to consider during classroom implementation. Some students also commented that the directions were easy to follow, allowing them to get a better grasp of the app despite having never tried it beforehand.

Another point to note is the students' responses when asked about their preferred device. None of the participants preferred the use of the smartwatch, mostly citing the larger screen size of other devices as a reason. Should this project continue to be deployed on a smartwatch, some care must be taken in the design of the interface so that the screen size is not too much of a point of contention due to the limited space it provides.

5 Conclusion and Future Work

This chapter presented the evaluation of *Estimate It!* as part of a larger study about the design and implementation of embodied cognition of mathematics educational systems.

Originally designed to run on Arduino Uno Microcontroller Cyberlearning Watches (Arroyo et al., in press), *Estimate It!* has now been successfully migrated to both AndroidOS and TizenOS. Specifically, the game can now be played on phones running on Android 2.3 (Gingerbread) or higher and on watches running on Tizen 2.3.1. The phone version is now capable of playing the game via NFC. Server code has also been refactored, deployed, and tested to run in a public domain.

Evaluation objectives were two-fold. This study aimed to investigate how the game compared with traditional instruction methods and what cultural, social and intellectual characteristics made the application suited (or not) to the teaching of estimation. Math teachers were invited to appraise the game's value in terms of the objectives. Most of the experts gave positive evaluations of the game.

In general, the evaluator teachers liked the collaborative nature of the technology and enjoyed how the game allows students to interact both with their classmates and their environment. There was appreciation for how the overall teaching method provided a more concrete and hands-on way of presenting estimation concepts as well as how the method allowed the teacher to monitor and move around the classroom together with the students, allowing for active participation and communication between students and teachers and among students themselves. The ability to include common household objects in the game had its appeal as it made the game seem accessible and easily customizable. The evaluators estimated that the game's competitive nature would make the method *more fun and exciting* than the usual classroom lecture. It was a common guess among the evaluator teachers that the injection of technology would be seen by students as attractive and motivating to them. These strong points made *Estimate It!* seem like a promising aid for the teaching of estimation concepts as part of the curriculum.

At the same time, there were areas of concern raised by the evaluators that the authors of this study consider *cultural* concerns. There was a common worry that implementation of the game in schools might be hindered because schools may not be financially ready for the required technology infrastructure. In particular, the need for a stable Internet connection was repeatedly mentioned. Acquiring devices (smartwatches or smartphones) for students to use could already be difficult. Other evaluators worried that setting up the game might be too tedious for some teachers, while one particularly disliked how playing the game removed the need for students to write down their mathematical working. One evaluator urged the researchers to reflect on how the game could accommodate classes of 40 students. Evaluators also dwelled on the propensity of students to misbehave while playing and on the possibility of students going off-task because of other interesting applications on the device.

In summary, evaluators seemed open to the idea of adopting a game-based reinforcement in their math classes. As one respondent put it, *when executed well, the advantages of* Estimate It! *will outweigh the disadvantages*.

As for the game's feasibility within a classroom setting, the application is still not ready to be deployed in its current state. From the student evaluation, it is possible that issues such as network connectivity will hinder students from efficiently completing tasks and cause large delays, making the game go on longer than it should be played.

Lastly, regarding its ease of use from the point of view of the students, the network connectivity once again might prevent students from completing the tasks. Other issues reported by and observed from students included small buttons, difficult-to-navigate screens, and slow responsiveness. Moreover, while errors in button pressing were not as frequent as previously feared, some improvements to the interface can still be made, as per the suggestions of the participants. However, the behavior of students indicated that they were able to engage with the application and not be completely frustrated by the issues they faced. Participants also reported that the game was generally fun, though they would rather play on a device with a larger screen such as a smartphone or a computer.

Moving forward, an iteration of the game that takes into consideration the findings of this study may be created and retested with evaluators and students to assess how they respond to this new way of teaching estimation. Given that previous evaluations had respondents that came from the United States, perhaps an evaluation of how this new method may affect the performance and understanding of estimation of Filipino students can be conducted. Results can then be compared to discover similarities and differences.

Acknowledgements We would like to thank the teacher experts and students for responding to the invitation to evaluate the game. We thank the Ateneo Laboratory for the Learning Sciences (ALLS) for the acquisition of items needed for the migration and the experiments (Samsung Gear S2, respondent tokens, toys and NFC tags etc.). We thank the Department of Information Systems and Computer Science (DISCS) for allowing us to use computer laboratories in order to conduct the evaluation.

References

Android NFC and Mobile Development Stacks for Web. Accessed May 20, 2016 from https://developer.android.com/reference/android/nfc/package-summary.html.

Esteves, A. (2012). Designing tangible interaction for embodied facilitation. In *Proceedings of the Sixth International Conference on Tangible, Embedded and Embodied Interaction* (pp. 395–396). ACM.

Esteves, A., van den Hoven, E., & Oakley, I. (2013, February). Physical games or digital games?: Comparing support for mental projection in tangible and virtual representations of a problem-solving task. In *Proceedings of the 7th International Conference on Tangible, Embedded and Embodied Interaction* (pp. 167–174). ACM.

Hornecker, E., & Buur, J. (2001). Getting a grip on tangible interaction: a framework on physical space and social interaction. In *Proceedings of the SIGCHI Conference on Human Factors in Computing Systems* (pp. 437–446). ACM.

Link, T., Moeller, K., Huber, S., Fischer, U., & Nuerk, H. C. (2013). Walk the number line—An embodied training of numerical concepts. *Trends in Neuroscience and Education, 2*(2), 74–84.

Malinverni, L., & Burguès, N. P. (2015). The medium matters: the impact of full-body interaction on the socio-affective aspects of collaboration. In *Proceedings of the 14th International Conference on Interaction Design and Children* (pp. 89–98). ACM.

Redish, E. F., & Kuo, E. (2015). Language of physics, language of math: Disciplinary culture and dynamic epistemology. *Science & Education, 24*(5–6), 561–590.

Rountree, W. L. (2015). *Redesigning traditional children's games to teach number sense and reinforce measurement estimation skills using wearable technology* (Master's Thesis).

Schaper, M. M., Malinverni, L., & Pares, N. (2014). Participatory design methods to define educational goals for full-body interaction. In *Proceedings of the 11th Conference on Advances in Computer Entertainment Technology* (p. 50). ACM.

Tizen native application development tutorials and developers forum. Accessed Feb 1, 2016 from https://developer.tizen.org/forums/native-application-development/active.

Whitton, N. (2009). Learning and teaching with computer games in higher education. In Connally, T., Stansfield, M., & Boyle, L. (Eds.), *Game-based learning advancements for multi-sensory human computer interfaces: Techniques and effective practices. Information science reference*. New York: Hershey.

Wilson, R. A., & Foglia, L. (2011). *Embodied cognition*. Accessed April 26, 2016 from http://stanford.library.usyd.edu.au/entries/embodied-cognition/http://stanford.library.usyd.edu.au/entries/embodied-cognition/, http://stanford.library.usyd.edu.au/entries/embodied-cognition/, http://stanford.library.usyd.edu.au/entries/embodied-cognition/.

Virtual Reality Enzymes: An Interdisciplinary and International Project Towards an Inquiry-Based Pedagogy

Ryan Ba, Yuan Xie, Yuzhe Zhang, Siti Faatihah Binte Mohd Taib, Yiyu Cai, Zachary Walker, Zhong Chen, Sandra Tan, Ban Hoe Chow, Shi Min Lim, Dennis Pang, Sui Lin Goei, H. E. K. Matimba and Wouter van Joolingen

Abstract Education in Science, Technology, Engineering, and Mathematics is moving towards more inquiry-based and creativity-stimulating pedagogies. Part of a curriculum based on such pedagogies should be challenging learning activities that engage students in investigation. At the same time, it is imperative that such activities are developed and validated in collaboration with the teachers who will incorporate them in their lesson planning. In this project, educators, researchers, and developers from Singapore and the Netherlands are working closely to develop innovative tools that assist biology education. Model-based and virtual reality-enabled solutions are

---

R. Ba · Y. Xie · Y. Zhang · S. F. B. M. Taib · Y. Cai (✉)
Nanyang Technological University, Singapore, Singapore
e-mail: myycai@ntu.edu.sg

Z. Walker · Z. Chen
National Institute of Education, Singapore, Singapore

S. Tan
Hwa Chong Institution, Singapore, Singapore

B. H. Chow
River Valley High School, Singapore, Singapore

S. M. Lim
Nanyang Girls' School, Singapore, Singapore

D. Pang
Riverside Secondary School, Singapore, Singapore

S. L. Goei
Windesheim University of Applied Sciences, Zwolle, Netherlands

S. L. Goei
Free University of Amsterdam, Amsterdam, Netherlands

H. E. K. Matimba · W. van Joolingen
Freudenthal Institute, Utrecht University, Utrecht, Netherlands

Y. Cai et al. (eds.), *VR, Simulations and Serious Games for Education*, Gaming Media and Social Effects, https://doi.org/10.1007/978-981-13-2844-2_5

being studied through interdisciplinary and international collaboration among the project members from the two countries.

Keywords Virtual reality · Serious games · Biology · Enzymes
Inquiry-based pedagogy

1 Introduction

There is a strong research base providing support for the argument that technology enables the involvement of students in realistic scientific tasks, for example through the use of virtual reality (Cai et al. 2013; Tan and Waugh 2013; Chow and So 2011), simulations (Cai and Goei 2013; Cai et al. 2016; Rutten et al. 2012; Rutten et al. 2015; Shin 2002; van Joolingen and de Jong 2003), external data and laboratories (Cai 2011, 2013; van Joolingen et al. 2005), as well as modelling tools (Blikstein et al. 2005; Louca and Zacharia 2011). These tools allow for the creation of task environments in which realistic and authentic forms of inquiry are possible and within the grasp of students in secondary education. The beneficial effect of such realistic contexts can be linked to the potential of technology to make students' learning experiences more authentic and to increase their exposure, engagement and interaction. Authenticity is considered important because more authentic learning experiences may lead to higher cognitive fidelity, which in turn could contribute to better learning outcomes.

A major goal of science education is to develop students' skills and attitudes towards Science, Technology, Engineering, and Mathematics (STEM), including their views on the nature of science, the role of scientific representations and models, and the processes of scientific inquiry and the creativity involved, as well as students' personal attitudes and possible future careers in science or technology. The construction and evaluation of scientific models (Löhner et al. 2005; Minner et al. 2010) provide a means to offer students experience with scientific research on a small scale and to learn about the model-based nature of scientific knowledge, especially about the relation between models and reality.

In STEM, the learning issue can be defined at two levels. At a domain-generic level, students need to learn about the nature of scientific knowledge and the role of models in representation. Within a specific domain, for example molecular and cell biology that is studied in this project, the relation between models and reality is particularly apparent. It is important that students gain insight into both the explanatory powers of models and their limitations in understanding in this domain, and learn to see the added value of using multiple representations and multiple models.

2 Virtual Reality Enabled Teaching and Learning

Virtual reality (VR) is widely considered to be one of the most viable technology applications for use in education. A large number of research papers has been published over the last few years discussing VR in education (Freina and Ott 2015; Merchant et al. 2014; Chow and So 2011; Cai 2013). This chapter describes a project in which Virtual Reality is applied in molecular biology. The project is a joint effort between Singapore (Nanyang Technological University (NTU), Hwa Chong Institution and River Valley High School) and the Netherlands (Windesheim University of Applied Sciences and Utrecht University). Based on existing work on VR technology development for curriculum-based learning and teaching of molecular biology (Cai 2011, 2013), this interdisciplinary and international project, brings together researchers from the two countries collaborate in developing VR-enabled solutions to be used in innovative lessons for molecular biology. The goal is to promote education research through international partnership conducted to craft pedagogy for VR-enabled learning and teaching, so as to add to the knowledge and evidence already available on VR education, model-based learning, and teachers' competencies in building innovative lessons for VR-enabled learning. The ultimate aim is that students in the classroom will benefit from this research. A previous experimental study by the Singaporean team showed that students do receive benefits from the use of VR technology in molecular learning (Tan and Waugh 2013).

Last but not least, this project is an attempt to bring together educators, researchers, and students using a cross-cultural partnership. The objective is the development of effective and validated pedagogies for inquiry-based learning, as well as skills and attitudes in students for modeling and stimulating relevant for functioning in current complex and rapidly-changing 21st century working environments.

3 Modeling in Enzyme Biology

In the biology domain, textbooks typically depict molecular and cellular processes such as enzyme operation and protein synthesis with iconic representations of macromolecules (Figs. 1 and 2). While this representation enables students to obtain a global view of the processes involved, there are aspects that are neglected despite being important for deeper understanding (Figs. 3 and 4). For instance, apart from the 'lock and key' idea of enzymes that is involved in order for molecules to 'snap' into each other, the molecules themselves are dynamic structures and their movement within the cells adds to the dynamics. Whereas the textbook representation may give rise to the misconception that molecules display purposeful behavior, a representation that incorporates dynamics can give rise to a more accurate 'mechanistic way' of reasoning that is capable of explaining the effects of external factors such as temperature and pH value in the cell (Figs. 5 and 6). This project is interested to develop models and modeling environments in which students can create and play

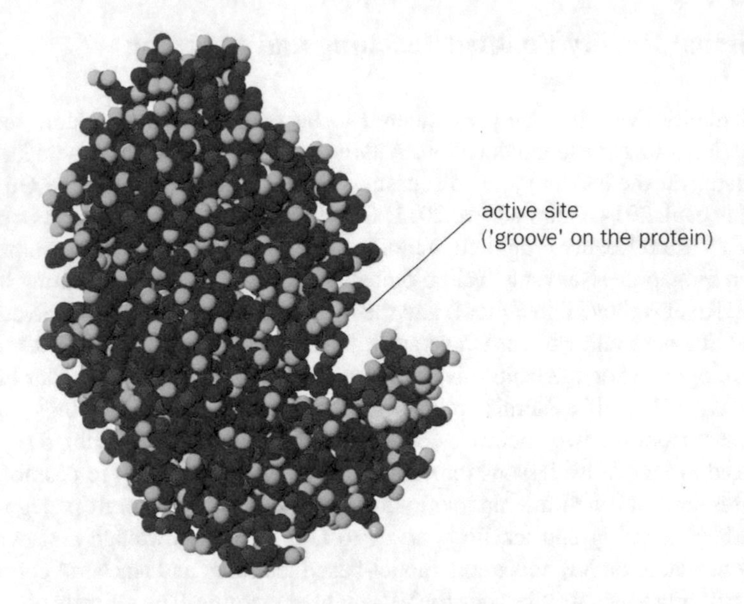

Fig. 1 A three-dimensional model of enzyme pepsin with its active site

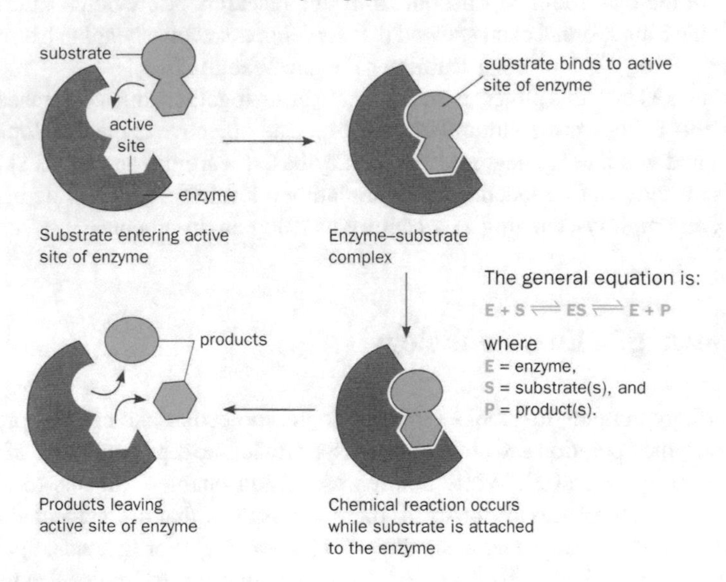

Fig. 2 The mode of action of an enzyme

with such multiple representations and use lesson plans to support learning in these environments. In this way, students can simultaneously learn concepts, processes, and functions within the domain of molecular and cell biology and develop 21st century skills related to the understanding of science and scientific knowledge.

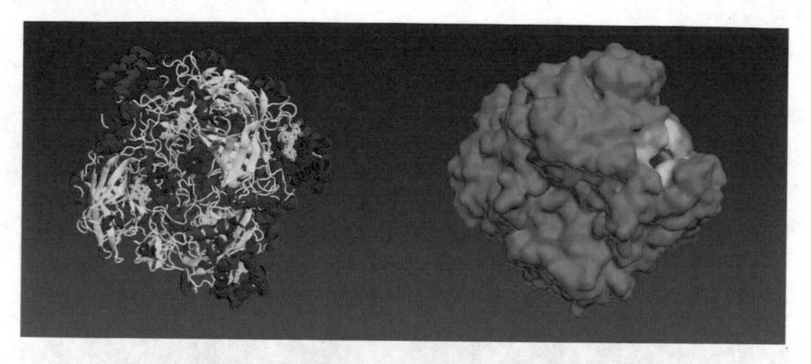

Fig. 3 The three-dimensional remodeling for VR application, 1DGB—human erythrocyte catalase

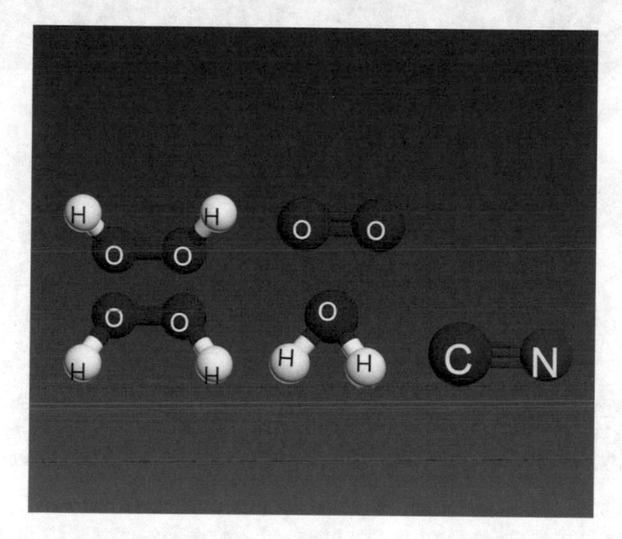

Fig. 4 The three-dimensional models of substrate molecules used in VR application. $2H_2O_2$, O_2, H_2O & CN−

4 VR Technology Enhanced Learning

Figure 7 shows students from the National Institute of Education doing VR enzyme learning in the virtual and augmented reality flipped classroom recently set up in Nanyang Technological University.

In the design, implementation, and evaluation of the modelling activities, teachers often actively involve the use of Lesson Study (Lewis et al. 2006). Lesson study is a collaboration-based teacher professional development approach that originated in Japan (Fernandez and Yoshida 2004). In this approach, teachers collaboratively engage in research inside their classrooms using a design cycle: preparing and designing lessons, performing the designed lessons as research lessons, and evaluating them in order to feed into the next cycle (Cerbin 2011). Teachers collaboratively design

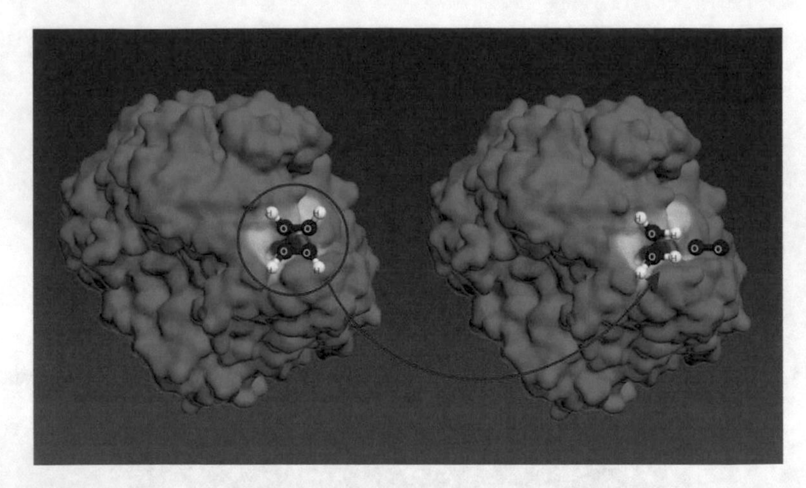

Fig. 5 *'Lock-and-Key' Hypothesis* model based interaction in VR Application

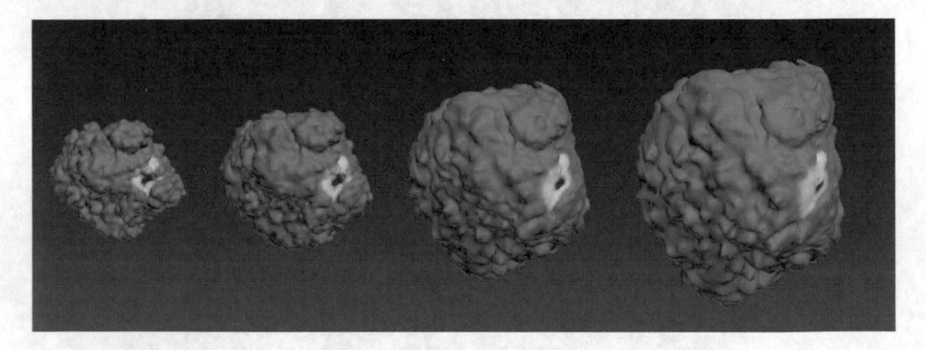

Fig. 6 Enzymes denaturation in VR application. Denaturation is the change in the three-dimensional structure of an enzyme or any other soluble protein, caused by heat or chemicals such as acids or alkalis. Enzymes lose their active sites when they are denatured

one or more research lessons in which they attempt to adjust to the varying educational and instructional needs of their students (Goei 2013). Great thought is devoted to predicting how the students may react. In the Lesson Study cycle, it is crucial for other teachers who are Lesson Study team members to observe the students during the research lesson and pay special attention to their learning activities and behavior. The lesson is evaluated immediately, with the focus being on the learning activity rather than the performance of the teacher who executed the lesson (Becker et al. 2008). Observations are shared, ways of refining and improving the lesson plan are discussed, and a subsequent review of the lesson is planned. In most cases, the adapted research lesson is then used for another class.

The project has now led to a set of lesson scenarios, developed in collaboration with teachers in both countries. The basic idea is to focus on the process of digestion in a story-like manner and have students investigate the factors that influence digestion.

Fig. 7 Interactive learning of molecules modelled

Fig. 8 The poster triggering the exploration of digestion. All the food items are active zones. Gazing at them or pointing to it with the mobile app will trigger a 3D exploration of the digestion of that item

The story starts with a poster of a woman who can choose from a number of available foods. When gazing at a particular kind of food, an animation is started showing the woman eating that food. After that, the student can zoom into the mouth, the stomach or the intestines and see and explore the working of the enzymes, in particular Amylase, which breaks down the starch, present in bread, rice, potatoes etc., to shorter chains of glucose elements. It is visualized that the enzyme will fit on starch, but not on cellulose, which is made up of the same glucose elements, but connected in a different way. Also students can explore how the enzymes behave under different conditions, such as changing temperature and variance in pH (acidity) of the environment (Fig. 8).

In this international collaboration, the Singaporean team has developed the simulations and visualizations for VR-enabled learning and teaching of enzymes whereas the Dutch team has developed the storyline around digestion and will integrate this into molecular biology lesson plans that employ the features of modeling and simulation. They will also validate and fine-tune the lesson plans via Lesson Study. The lesson plans will also be tested out by teachers in Singapore, and by subject matter specialists in the Netherlands.

The VR-enabled solutions draw on the work of the NTU team. They are highly specialized in biology modelling, visualization, interaction and user-interface. With the VR tool, students can deal with two-dimensional (2D) drawings, three-dimensional (3D) graphics of molecular structure, and four-dimensional (3D plus time) simulations of dynamic functions. The pedagogy is based on inquiry-based learning and modeling using the approach of modeling and simulation for learning. Focus in this approach is the integration of the inquiry experiences within the classroom, with the teacher orchestrating students explorations and reasoning. The novelty of the development is its embracing of more general concepts of models across multiple levels and with multiple variables for teaching and learning. The models include simulation and 2D and 3D visualizations. The value of Lesson Study in this process is that the designed pedagogies and lesson plans will take the roles of teachers and students into account. The detailed predictions of student behavior and the classroom observations will provide insight in the reasoning patterns that will be triggered by the 3D visualizations. The Lesson Study stage of this project follows the model designed by De Vries et al. (2016) for the Netherlands. This model entails a structured way for lesson design, selection of case students, prediction of student behavior and lesson observation. Lesson Study is also currently part of professional development initiatives in Singapore schools.

5 Conclusion

This project is timely in its exploration of VR technology in education (especially STEM education), aiming to help pupils achieve better understanding of scientific concepts through model-based simulation for interactive learning. One of the existing challenges for higher education is internationalizing its programs and making students more globally competent. Although teaching is often tailored to local contexts, international collaboration is important for educators as they can grow professionally through exposure to innovative ideas and best practices in other settings. It may be expected that teaching traditions and learner and teacher dispositions are different in the two countries, which would provide all teachers involved with fresh insights about teaching and learning.

Acknowledgements This project is partially supported by Ministry of Education Academic Fund (Singapore part) and RAAK-SIA (Dutch part).

References

Becker, J., et al. (2008). A college lesson study in calculus, preliminary report. *International Journal of Mathematical Education in Science and Technology, 39,* 491–503.

Blikstein, P., et al. (2005). NetLogo: Where we are, where we're going. In M. Eisenberg & A. Eisenberg (Eds.), *Presented at the annual meeting of interaction design and children*. Boulder, Colorado.

Cai, Y. (Ed.). (2011). Interactive & digital media for education in virtual learning environment. Nova Sciences.

Cai, Y. (Ed.). (2013). 3D immersive & interactive learning. Berlin: Springer.

Cai, Y., & Goei, S. L. (Eds.). (2013). *Simulation, serious games and their applications*. Berlin: Springer.

Cai, Y., Goei, S. L., & Trooster, W. (Eds.) (2016). Simulation and serious games in education. Berlin: Springer.

Cai, Y., et al. (2013). Introduction to 3D immersive and interactive learning. In Y. Cai (Ed.), *3d immersive and interactive learning*. Berlin: Springer.

Cerbin, B. (2011). *Lesson Study: Using classroom inquiry to improve teaching and learning in higher education*. Sterling: Stylus Publishing.

Chow, B. H., & So, K. L. (2011). The VR Classroom @ River Valley High School. In Y. Cai (Ed.), *Interactive and digital media for education in virtual learning environments*. New York: Nova Sciences, May 2011, pp. 129–139.

De Vries, S., Verhoef, N. C., & Goei, S. L. (2016). *Lesson Study: een praktische gids voor het onderwijs*. Garant: Apeldoorn/Antwerpen.

Fernandez, C., & Yoshida, M. (2004). *Lesson Study: A Japanese approach to improving mathematics teaching and learning*. Mahwah: Erlbaum.

Freina, L., & Ott, M. (2015). A literature review on immersive virtual reality in education: State of the art and perspectives, eLearning and Software for Education Conference, April 2015, Romania.

Goei, S. L. (2013). Lesson study as a professional tool to strengthen teachers' competencies in designing educational activities for differential (additional) educational support needs. Invited Plenary for the World Association Lesson Study conference 2013, September 5–8, Gothenburg, Sweden.

Lewis, C., et al. (2006). How should research contribute to instructional improvement? *Educational Researcher, 35,* 3–14.

Löhner, S., et al. (2005). Student's reasoning during modeling in an inquiry learning environment. *Computers in Human Behavior, 21*(3), 441–461.

Louca, L. T., & Zacharia, Z. C. (2011). Modeling-based learning in science education: cognitive, metacognitive, social, material and epistemological contributions. *Educational Review*, 1–22.

Merchant, Z., Goetz, E. T., Cifuentes, L., Keeney-Kennicutt, W., & Davis, T. J. (2014). Effectiveness of virtual reality-based instruction on students' learning outcomes in K-12 and higher education: A metaanalysis. *Computers & Education, 70,* 29–40.

Minner, D. D., et al. (2010). Inquiry-based science instruction-what is it and does it matter? Results from a research synthesis years 1984 to 2002. *Journal of Research in Science Teaching, 47*(4), 474–496.

Rutten, N., van Joolingen, W. R. & Van der Veen, J. T. (2012). The learning effects of computer simulations in science education. *Computers & Education, 58*(1), 136–153.

Rutten, N., van der Veen, J. T., & van Joolingen, W. R. (2015). Inquiry-based whole-class teaching with computer simulations in physics. *International Journal of Science Education*, 1–21.

Shin. Y. (2002). VR simulations in web-based science education. Computer App. in Eng. Education 10(1), 18–25

Tan, S., & Waugh, R. (2013). Use of virtual-reality in teaching and learning molecular biology. In Y. Cai (Ed.), *3D Immersive and Interactive Learning,* Springer 2013, pp. 17–43.

van Joolingen, W., & de Jong, T. (2003). SimQuest: Authoring educational simulations. In T. Murray, S. Blessing, & S. Ainsworth (Eds.), *Authoring tools for advanced technology educational software: Toward cost-effective production of adaptive, interactive, and intelligent educational software*. Dordrecht: Kluwer Academic Publishers.

van Joolingen, W., de Jong, T., Lazonder, A. W., Savelsbergh, E. R., & Manlove, S. (2005). Co-Lab: Research and development of an online learning environment for collaborative scientific discovery learning. *Computers in Human Behavior, 21*(4), 671–688.

Motion Sensing Games for Children with Autism Spectrum Disorder

Chenyi Wu and Quan Zheng

Abstract The use of motion sensing games is a new approach in autism education in mainland China. Currently, a number of challenges in its practical application remain. For instance, motion sensing games are still in their infancy, with few educational rehabilitation and practical skills games available. Assessment function with such games is not common and generalization of the game-based learning is difficult as well. To solve these difficulties and to improve the efficiency of motion sensing game applications, Suzhou Industrial Park Ren Ai School has cooperated with commercial enterprises, improved game design, enhanced the role of assessment, and incorporated pivotal response training in order to meeting all of the challenges.

Keywords Motion sensing · Game · Autism education practice

1 Introduction

The educational practice and application of information and communications technology (ICT) has been regarded as a vital trend of ICT development in China. Guidelines issued by the Ministry of Education of the People's Republic of China (2010) of the National Medium-and-Long-Term Educational Reform and Development (2010–2020) and the plans for education reform and development of different provinces and cities of China have fully confirmed the revolutionary influence of ICT on education. For children with autism, the application of ICT would enhance their comprehension skills, including social communication, emotional expression, and sensory integration. Motion sensing games based on virtual reality technology are thus a great innovation in autism education application.

Motion sensing games are a new type of interactive electronic game that operate by changing body actions (Liu et al. 2015). In 2006, the game company Nintendo released Wii Remote (2006) which marked the official global launch of the first ever

C. Wu (✉) · Q. Zheng
Suzhou Industrial Park Ren Ai School, Jiangsu Province, People's Republic of China
e-mail: 385809232@qq.com

Y. Cai et al. (eds.), *VR, Simulations and Serious Games for Education*, Gaming Media and Social Effects, https://doi.org/10.1007/978-981-13-2844-2_6

motion sensing game. In March 2010, SONY released PlayStation Move. At the time, wireless handhold equipment (2010) became a highlight in the world of video games. In October 2012, Microsoft released the Kinect motion controller. With the user able to use hand gestures alone, Kinect stands out among other motion sensing devices. Motion sensing games based on Kinect technology became the most popular games. Kinect has a powerful skeletal tracking function which can track two complete skeletons, each with 20 articulations (Han et al. 2013), makes an accurate assessment of different postures, and identifies all the user movements and postures in the game and immediately provide feedback to the player. In the field of special education in China, motion sensing games have been widely applied in sensory integration training (Wang et al. 2016), medical rehabilitation (Wang et al. 2014), parent-child interaction (Ma et al. 2012), and physical training (Li et al. 2012), among others. At present, Kinect has been found to be one of the best technological resources for autism education.

In Suzhou Industrial Park (SIP) Ren Ai School, wide-ranging applications of motion sensing games in autism education are being studied. Firstly, motion sensing games can reduce behavior problems and help children with autism spectrum disorders (ASD) adapt to different environments. Emotional behavior disorders are one of the factors impeding autistic children's socialization. Emotion is defined as the psychological experience and mental state produced when people receive environmental stimulation. Most autistic children have difficulties in adapting themselves to the environmental conditions, which leads to emotional behavior disorders. Motion sensing games can create varieties of three-dimensional (3D) virtual reality situations according to user requirements (Han et al. 2013), providing autistic children with the most adaptive and comfortable training environment. After the training, the incidence of emotion behavior disorders is greatly reduced among autistic children (Ge and Fan 2017). In addition, autistic children pay more attention during game-based training to pursuing the feeling of success in the game and pay less attention to the environment around them, and feel comfortable about adapting themselves to the gradual changing environment. At the end of the game, autistic children leave the virtual reality and return to the real world. Some of them do not feel comfortable while others adapt well as they have already experienced many different environments in the virtual reality motion sensing game. After a two-year on-track study, SIP Ren Ai School's research team came to the same conclusion: The motion sensing game based on virtual reality significantly improved autistic children's emotional regulation and joint attention (Ge and Fan 2017). And in October 2014, the SIP Ren Ai School presented a paper on Social Development for Children with Autism Using Kinect Gesture Games: A Case Study in Suzhou Industrial Park Ren Ai School at the Netherlands Asia–Europe International Forum that had a theme of 3D Virtual Technology Application in Educational Practice.

Secondly, social communication is another big challenge for autistic children. They often do not know how to socialize under many circumstances. Though they may have developed some skills in social communication, these skills are unfortunately used only for particular purposes as a tool (Xiong 2012). Children with autism do not understand the enjoyment of sharing experiences through social communication. Motion sensing games may help them understand various kinds of human-computer interaction through verbal and body language as well as facial expression and develop simple social skills. At the same time, multiplayer games that require communication and cooperation among players might arouse children's interest in social contact based on the sharing of personal experiences and further improve their social communication abilities. Some research has shown that motion sensing games can improve autistic children's attitudes towards peer interactions and increase the frequency of their social interactions (Bartoli et al. 2014; Schreibman 2007; Bianchi-Berthouze 2013).

Further, Oregon State University researchers have found a relationship between motor skill deficiencies and the severity of autism spectrum disorder symptoms in young children (Macdonald et al. 2013). A study was conducted on the development and motor skills of 110 children with an autism diagnosis who were aged 12–33 months old. The results showed that children with weaker motor skills had greater social communicative skill deficits. A follow-up by the same authors showed that developing motor skills was essential for the development of children with autism spectrum disorders and could help develop better social skills (Macdonald et al. 2014). Li et al. (2012) research points out that motion sensing games can effectively promote the development of motor skills such as body coordination, muscle strength, and endurance in autistic children.

The Nottingham Institute of Education at Nottingham Trent University aimed to enhance special needs children's learning interest and learning ability with the use of athletic sports motion sensing games. As a result, 92% students preferred using electronic games to improve their studies, 93% students agreed that the game had increased their learning interest, and 87% students stated that they had learned something they had never thought of before (Kandroudi and Bratitsis 2012).

2 Challenges of Motion Sensing Games in Autism Education

2.1 Needs of Local Motion Sensing Games

Nearly 200 Kinect-based motion sensing games are available on the XBOX platform, including athletic sports, formation, role playing, fighting, music, and casual puzzle games. However, statistics show that 87.7% of the motion sensing games available on the XBOX platform in mainland China are in English while the remainder use traditional Chinese instead of simplified Chinese. Moreover, only 17.4% of the traditional Chinese versions actually use Chinese dialogues instead of English, French, or Japanese speech. Therefore, motion sensing game classes for autistic children are often interrupted due to translation problems, to negative effect.

2.2 Lack of Game Resources on Education

As Kinect is an open game platform and games are available to wide audiences, each XBOX game is assigned an age appropriacy level according to the Entertainment Software Rating Board (ESRB) rating system before its release (Haninger and Thompson 2004). Through references to each game's classification in the ESRB and rigorous testing by the trainer, 19 games (out of the 187 available motion sensing games at the time) were selected as being suitable for autistic children education and rehabilitation. Few games are available for autistic children specifically, thus decreasing the variety of virtual reality situations available and appropriate for use in this study. The lack of suitable games also affects the children's interest and persistence in training.

2.3 Low Applicability of Game

Most motion sensing games based on Kinect or the XBOX platform are developed by commercial game companies who do not consider the needs of special children, especially autistic children, as it is a very niche market. Although there are a considerable number of motion sensing games which have exquisite life-like pictures, the pace of in-game pictures and voice instructions is often too rapid for autistic children, who struggle to keep up. Very often, participants in this study stood still in front of the game screen—even though they were fascinated by the games, they found the games difficult to follow. Thus, commercial motion sensing games cannot help to substantially boost the rehabilitation of autistic children. Although there is a small number of appropriate motion sensing games (based on independent research and development) available on the market, there are too few games to be able to fulfil the sustained, long-term need that rehabilitation training for autistic children poses.

2.4 Poor Assessment of Motion Sensing Games' Application Efficiency

As motion sensing games originated outside of China, no standard training program based on motion sensing games has yet been developed. Very few studies have focused on the application of motion sensing games in autism education and it is difficult to make an assessment of the effects of any specific motion sensing game. For example, in one study by Xu and Ji (2016), the Pediatric Evaluation of Disability Inventory was used for participant assessment, while the Psycho-educational Profile—Third Edition was used to assess participants' social development in a different study (Ge and Fan 2017). These frameworks are based on basic physiological and psychological assessments and do not measure the academic abilities and extent of rehabilitation of children with autism. Due to the lack of an assessment systems, teachers have no criteria for setting training program goals for children with autism and cannot ascertain if goals have been met. Another factor that contributes to the reduced effectiveness of motion sensing game training is the causing of sports injuries through excessively difficult games (Li and Gao 2017).

2.5 *Difficulties in Skill Generalization*

As we know, the environment in motion sensing game is virtual. However, there is a gap between the experience and skills acquired in the virtual game and real life (Fu 2017), as certain skills acquired by autistic children in virtual reality may not be applicable to the real world. Virtual reality is preset and does not change for a long time, while the real world is full of changes and uncertain factors that might affect autistic children and directly hamper the utilization of their acquired skills. Thus, the implementation and generalizing of skills acquired by autistic children in virtual reality to the real world remains significantly problematic.

3 Applications of Motion Sensing Game Application in Autism Education

3.1 *Strengthen Collaboration Between Schools and Commercial Enterprises to Promote Game Research and Development*

It is difficult for special educational schools to independently develop motion sensing games as teachers usually do not have computer programming skills or backgrounds. The same is true for commercial enterprises who are often unfamiliar with the needs of autistic children. Special educational schools should thus collaborate with commercial enterprises to bring about personalized motion sensing games for autistic children.

A good example of one such collaboration is the Pink-Dolphin Program in which a game was developed jointly by Ren Ai School, Nanyang Technological University in Singapore, and Windesheim University of Applied Sciences in the Netherlands. The game has been shown to significantly improve autistic children's sensory integration, communication motivation, and focusing ability, among others (Cai et al. 2017). Figure 1 shows a child with ASD playing the Pink-Dolphin motion sensing game.

Fig. 1 Child with ASD playing the Pink-Dolphin motion sensing game

In collaboration, the research teams have successfully produced five virtual reality motion sensing games based on Kinect. These games have been modified to meet the specific needs of students in Ren Ai School (Cai et al. 2013) and used in autism education classes. See Table 1 for descriptions of the five motion sensing games.

3.2 *Improving the Assessment Process and Implementing Efficient Training*

Assessment is indispensable in the education and rehabilitation of autistic children, but not every teacher of autistic children is professionally qualified to do so. Therefore, professional assessment teams are desperately needed in special educational schools. These teams would be committed to task assessment and be responsible for turning the assessment results into dynamic descriptive text or graphics so as to ensure effective dissemination of training results. For example, in a study of the effect of motion sensing game interventions on children with autism, teachers at Ren Ai School obtained Figs. 2 and 3 by observing and evaluating the success rate of students' target behavior.

With graphs, readers intuitively understand that the reversal design study found that the Pink-Dolphin motion sensing game improved the upper limb movement ability and action simulation ability of children with autism.

Table 1 The five motion sensing games

Name	Label	Description	Aim	Difficulty	Interest
Pink-Dolphin	ACT (Action game)	The player is a little dolphin trainer in a virtual dolphin lagoon. To make the dolphins happy, the player must make specific actions as prompted. The more accurately the actions are performed, the higher the player's score and the happier the little dolphins become	• Improve sensory integration ability of children with autism • Express emotion correctly and stimulate students' willingness to communicate • Improve the stability of autistic students' upper limbs movements • Improve autistic students' joint attention	★	★★★
Shopping	SIM (Simulation game)	In a virtual supermarket, players push carts to buy commodities. Each commodity has a price, and players reach out with their hands to put items into their cart. After selection, the player goes to the cashier and waits in a queue to pay	• Practice mathematical skills, including understanding numbers and quantity, mathematical calculation, recognizing coins, and using a calculator • Comparison and selection • Acquire the social rules of supermarket shopping • Express emotion correctly	★★★	★
Cross the Road Safely	PZL (Puzzle game)	On a virtual busy road, players observe traffic lights and move forward on a zebra crossing to cross a road successfully	• Know a road, cars, people, sidewalks, traffic lights • Know the role of traffic lights • Cross a road according to rules	★★★	★★
I Love Taking a Bath	RPG (Role-playing game)	In a virtual bathroom, "your" (the player's) hair is disheveled, you are physically dirty, and you smell. You need to turn on the tap and select the correct shampoo to wash your hair and the correct shower gel to clean your body. You use your hands to scrub until you are clean	• Reduce fear of water in some autistic children. Allow them realize that "bathing is fun" • Acquire the general process of bathing • Be able to distinguish shower gel from shampoo and know each item's function • Express emotion correctly	★★	★★

(continued)

Table 1 (continued)

Name	Label	Description	Aim	Difficulty	Interest
Penalty Shootout	SPG (Sports game)	Members of two teams have a penalty shootout to determine the winner. Their eyes must stare at the ball and they must aim at the goal and kick with one foot	• Comply with the rules of the game • Play the game by changing the positions of the goalkeeper and the ball, improving students' joint attention • Improve students' hand-eye coordination • Improve students' social skills in the competition	★★	★★★

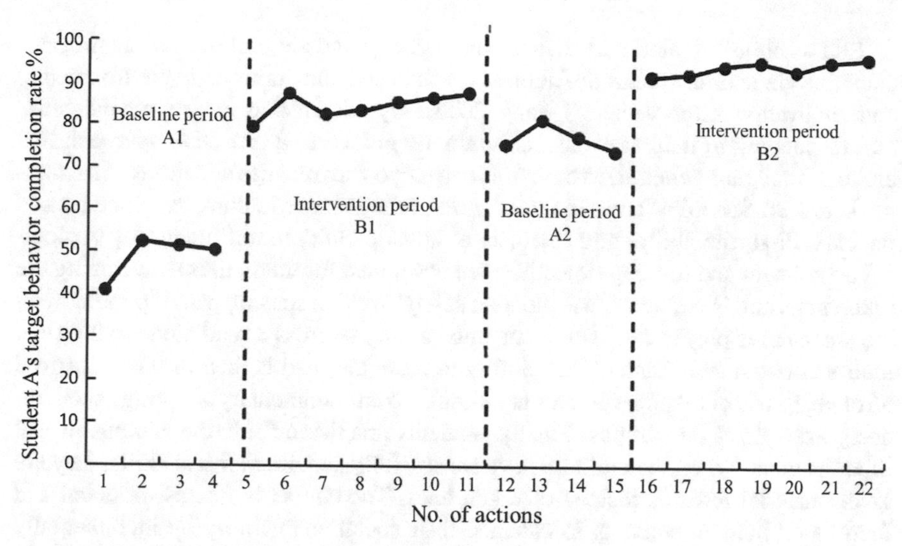

Fig. 2 Student A's target behavior completion rate

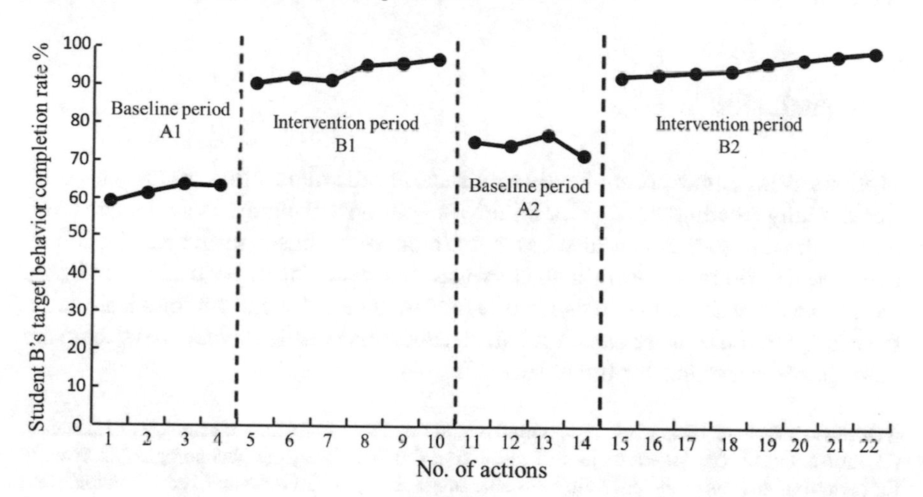

Fig. 3 Student B's target behavior completion rate

3.3 Adopting Pivotal Response Training and Generalizing Skills in the Natural Environment

Motion sensing game training's focus on specific, repeated movements is highly effective in helping autistic children to acquire certain practical skills. However, once the environment changes, it might be difficult for the children to show the skills acquired in virtual reality due to their stereotyped thinking patterns. Pivotal response training (PRT) is one of the effective ways to solve the problem.

PRT is child-oriented and focuses on assisting autistic children to obtain critical skills via natural situations (domestic, scholastic and community environments) and motivation improvement (Wang 2013). By identifying environmental clues and responding to natural events, children are guided to a proactive approach that enhances and thus generalizes their ability to cope with real life situations. An example is Ren Ai School's "cross the road" non-traditional curriculum, developed based on PRT. First, the ability and interests of autistic children are judged by ecological assessment and training objectives are developed for them, thus determining the token program. Next, the "cross the road safely" motion sensing game (popular with the students) is played to develop students' ability to cross a road alone in a virtual reality environment. Each child's ability to cross the road is then further improved through the use of token incentives in a simulated environment by adopting specially-designed courses on campus. Finally, students practice outside the campus in real life situations. When the children have successfully acquired critical skills, they are given material rewards instead of token ones. The course is interest-oriented and focuses on helping students to enhance their social adaptability by incrementally applying the critical skills they acquired in virtual reality to the real world.

4 Conclusion

Motion sensing games are eminently applicable in autism education around the world. Besides fully meeting the need for intuitive cognition that children with autism have, motion sensing games can also assist children with autism in the acquisition of experiences. Motion sensing games explore the potential of each child and help them to increase and improve their skills for daily life and social communication. As the use of motion sensing games in autism education is still relatively new, there are many possible avenues for future research.

Acknowledgements This work is supported by the Key Project of Chinese Ministry of Education (Grant No. DLA150334). We thank Professor Yiyu Cai from Nanyang Technological University for reviewing this paper and providing valuable suggestions, Li Fan for helpful conversations, and Qunchao Zhang for producing several of the figures.

References

Bartoli, L., Garzotto, F., Gelsomini, M., Oliveto, L., & Valoriani, M. (2014). Designing and evaluating touchless playful interaction for ASD children. In *Proceedings of the 2014 conference on Interaction design and children*: 17–26.

Bianchi-Berthouze, N. (2013). Understanding the role of body movement in player engagement. *Humana computer Interaction, 28*(1), 40–75.

Cai, Y., Chia, K., Thalmann, D., Kee, N. K., Zheng, J., & Thalmann, N. M. (2013). Design and development of a virtual dolphinarium for children with autism. *IEEE Transactions on Neural Systems and Rehabilitation Engineering, 21*(2), 208–217.

Cai, Y., Chiew, R., Nay, Z. T., Indhumathi, C., & Huang, L. (2017). Design and development of VR learning environments for children with ASD. *Interactive Learning Environments, 25*(6), 1–12.
Fu, H. (2017). Swot-pest analysis of introducing sports somatosensory games into school physical education class. *Journal of Youth Sports, 2,* 75–77.
Ge, Z., & Fan, L. (2017). *Social development for children with autism using Kinect gesture games: A case study in Suzhou Industrial Park Ren Ai School*. Springer Singapore: Simulation and Serious Games for Education.
Han, J., Shao, L., Xu, D., & Shotton, J. (2013). Enhanced computer vision with Microsoft Kinect sensor: A review. *IEEE Transactions on Cybernetics, 43*(5), 1318–1334. In Chinese.
Haninger, K., & Thompson, K. M. (2004). Content and ratings of teen-rated video games. *JAMA, 291*(7), 856–865.
Kandroudi, M., & Bratitsis, T. (2012). Exploring the educational perspectives of XBOX Kinect based video games. *6th European Conference on Games Based Learning*: 219.
Li, J., & Gao, Y. (2017). The advantages and disadvantages of somatosensory games on the development of preschool children. *Journal of Educational Development, 1,* 45–48.
Li, Lou, K. H., Tsai, S. J., Shih, H. Yin., & Ru-Chu. (2012). The effects of applying game-based learning to webcam motion sensor games for autistic students' sensory integration training. *Turkish Online Journal of Educational Technology, 11*(4), 451–459.
Liu, Y., Lei, X., & Hu, X. (2015). A review of foreign research into interventions in Autistic children by using somatic games and its enlightenment. *Chinese Journal of Special Education* (5) (in Chinese).
Macdonald, M., Lord, C., & Ulrich, D. A. (2013). The relationship of motor skills and social communicative skills in school-aged children with autism spectrum disorder. *Adapted Physical Activity Quarterly, 30*(3), 271.
Macdonald, M., Lord, C., & Ulrich, D. A. (2014). Motor skills and calibrated autism severity in young children with autism spectrum disorder. *Adapted Physical Activity Quarterly, 31*(2), 95–105.
Ma, J., Zhang, S., & Li, F. (2012). Design and implementation of parent-child interactive game based on somatosensory technology. *China Educational Technology, 9,* 85–88.
Ministry of Education of the People's Republic of China. (2010). Retrieved April 27, 2018, from http://www.moe.gov.cn/jyb_xwfb/s6052/moe_838/201008/t20100802_93704.html.
PlayStation Move. (2010). Retrieved April 27, 2018, from https://asia.playstation.com/move/tc/.
Schreibman, L. E. (2007). The science and fiction of autism. *Library Journal, 43*(3), 559–563.
Wang, Z. (2013). The new use and reflection of PRT intervention model for autism. *Journal of Sichuan University of Arts and Science, 23*(5), 80–83.
Wang, Y., Liang, J., Wan, Q., Huang, Z., & Gao. (2016). Design and implementation of motion sensing game based on somatosensory technology. *China Educational Technology & Equipment*, (2), 53–57 (in Chinese).
Wang, J., Ma, J., Chen, C., & Zhang, W. (2014). Effect of motion sensing games Kinect on executive function of stroke patients. *Chinese Journal of Rehabilitation Medicine, 29*(8), 748–751.
Wii Remote. (2006). Retrieved April 27, 2018, from https://www.nintendo.co.jp/wii/features/wii_remote.html.
Xiong, X. (2012). The construction of "Experience sharing" and its application in the rehabilitation of Autistic children. *A Journal of Modern Special Education, 1,* 55–57.
Xu, Y., & Ji, L. (2016). Effects of motion sensing games on children with Autism. *Chinese Journal of Clinical Psychology, 24*(4), 762–765.

Vehicle Behaviors Simulation Technology Based on Neural Network

Xin Yang, Shuai Li, Baocai Yin, Qiang Zhang, Guozhen Tan, Dongsheng Zhou and Xiaopeng Wei

Abstract In order to enhance the realism and diversity of traffic flow modelling, this chapter presents a data-driven traffic behavior model based on neural networks in a real-virtual interaction traffic simulation system. First, we extract individual personalized real trajectories from each vehicle, then use neural networks to develop specific traffic models from the trajectories of each vehicle. In contrast to traditional, manually-defined traffic models, we aim to develop a data-driven model to describe the relationship between the traffic states faced by a driver and the driver's resultant actions. In this model, a driver's behavior is influenced by the current traffic states of the leading vehicle and the following vehicle. This is a regression problem for which the inputs of the model are the traffic states of the leading and following vehicles. The output is the action of the current vehicle. Finally, this chapter presents a real-virtual interaction system. In detail, real trajectories are introduced directly into the simulation process to maximize the characteristics of real traffic flow. In comparison

X. Yang (✉) · B. Yin · Q. Zhang · G. Tan · X. Wei
Dalian University of Technology, Dalian, China
e-mail: xinyang@dlut.edu.cn

B. Yin
e-mail: ybc@dlut.edu.cn

Q. Zhang
e-mail: zhangq@dlut.edu.cn

G. Tan
e-mail: gztan@dlut.edu.cn

X. Wei
e-mail: xpwei@dlut.edu.cn

S. Li
Hong Kong Polytechnic University, Hung Hom, China
e-mail: lishuai9401@gmail.com

D. Zhou
Dalian University, Dalian, China
e-mail: donyson@126.com

Y. Cai et al. (eds.), *VR, Simulations and Serious Games for Education*, Gaming Media and Social Effects, https://doi.org/10.1007/978-981-13-2844-2_7

to existing simulation methods, traffic flows simulated by this method can depict irregular vehicle driving behavior.

Keywords Neural network · Traffic simulation · Microscopic model

1 Introduction

1.1 Background

The effectiveness of the transport system is crucial for the modern industrialized society. Rapid industrialization has led to rapid growth in the number of vehicles, bringing many challenges such as traffic congestion and the need to improve traffic network design and signal optimization. These challenges have had an impact on the economy, energy usage, and the environment. It is obvious that these problems have become a global issue. Fortunately, traffic simulation has become an effective solution to these problems, with the help of computer technology. Moreover, with the help of visualization technology, traffic simulation also has great applications in the entertainment industry. With the massive amount of data provided by advanced technology equipment such as the Global Positioning System (GPS) and embedded sensors, neural networks are able to develop driver models that produce much better results in traffic optimization than traditional models.

1.2 Related Research

This section contains a brief introduction of the different classifications of traffic models as well as an introduction to the traditional traffic flow simulation method.

In general, traffic flow simulation models can be divided into three categories: micro-simulation, macro-simulation, and medium-simulation.

1.2.1 Macro Model Based on Fluid Mechanics

The macro model tries to characterize the traffic flow with macroscopic quantities such as average density $\rho(x, t)$, average velocity $v(x, t)$ and flow rate $\theta(x, t)$.

Lightill and Whitham (1955) used a hydrodynamic model (usually referred to as the LWR model) to simulate traffic flow. The LWR model uses first-order equations in different spaces to calculate the state of each part of the vehicle flow. Since then, Newell (1993), Daganzo (1995), and Lebacque (1997) have all extended this theory based on Lightill's work. Sewall et al. (2010) has further proposed a continuum-based model that can be used to simulate real traffic flow on a large-scale road network.

The main drawback of macro-simulation is its lack of control over the behavior of the vehicles on an individual basis.

1.2.2 Micro-Model Based on Self-driven Particle Theory

Micro-simulation aims to model the behavior of each vehicle, hoping to use a series of complex rules to describe the vehicle's dynamic traffic behavior. Every vehicle in this model is an agent, and each agent is capable of responding to the traffic conditions of the surrounding vehicles and the surrounding traffic environment, and formulating driving strategies.

Many researchers have contributed to the progressive development of the micro-simulation model. Pipes (1953) first proposed a vehicle-following model in 1953. Gerlough (1955) then summarized these driving rules in a 1955 paper that was later adopted by a series of models, such as the nonlinear vehicle following model (Gazis et al. 1961), the cellular automata model (Nagel and Schreckenberg 1992), the optimization speed model (Bando et al. 1995), and the intelligent driving model (Treiber and Helbing 2001; Kesting et al. 1928). Shen and Jin (2012) proposed a detailed approach to rebuild urban transport networks. Kesting et al. (2007) introduced a lane change model called MOBIL in his 1997 paper and optimized the vehicle-following model. Hidas (2005) used an extended model to express driver behavior in the case of traffic congestion. Qiao (2008) analyzed the characteristics of traffic flow from traffic flow data and calibrated a number of parameters, including the driver reaction time in the vehicle-following model.

The key advantage of micro-simulation is its ability to show drivers' individual behaviors, and it has been applied to many traffic simulation platforms.

1.2.3 Medium-Model Based on Aerodynamics

The medium-model is a traffic model between the micro-model and the macro-model. The medium-model uses Boltzmann equations to simulate the dynamic process of traffic flow. This method was first proposed by Prigogine and Andrews (1960), with Nelson et al. (1997) later proving the model's effectiveness. In addition, Sewall et al. (2011) proposed a hybrid model that combines microscopic and macroscopic models and simulates large-scale traffic flows using real-time algorithms.

1.2.4 Data-Driven Behavioral Learning

There are some data-driven behavioral learning strategies in crowd simulation. Some methods develop a new population behavior model from the input video. For example, the data-driven approach used by Lee et al. (2007) trained group behavior models to match the behavior in real video. Ju et al. (2010) also proposed a method of combining existing population data to generate new crowd animation.

Bi et al. proposed a data-driven lane change model which divides the lane change process into a decision-making phase and an execution phase. The driver's behavior in both stages is determined by the model developed from the real data. During the decision-making phase, the author uses the random forest theory and the distance between the current car and the target lane ahead of and behind the car to determine whether to change the lane and consider the direction of lane change. In the execution phase, the paper presents a three-layer neural network model with the input as the traffic status of the current car and the output as the current car speed. Both stages of the model are developed from a large amount of real traffic data.

1.3 Drawback of Traditional Simulation Methods

The disadvantages of traditional simulation methods mainly exist in the following aspects:

(1) There is no concept of a real car, which means every car is an abstract entity. The traffic flow simulated by this method is relatively stable and thus does not reflect the changing traffic flow in real world.
(2) Traditional simulation methods use traditional microscopic models such as the intelligent driving model (IDM). The parameters involved in these models are considered to be empirically determined and cannot reflect the driving characteristics of drivers in the real world.

2 Related Theory and Technology

2.1 Intelligent Driven Model

The IDM is a widely used micro-model which assumes that each vehicle's behavior at the current moment is affected by the preceding vehicle. In the IDM, traffic conditions at a particular time are described by location, speed, and lane. The driver's decision to accelerate or decelerate depends on its current speed and the relative speed and distance to the vehicle in front. The vehicles behind make decisions based on the surrounding traffic environment. The purpose of the decision is to maintain a safe distance from the vehicle ahead and to achieve the desired driving speed. The IDM defines vehicle acceleration as a function of speed, distance, and relative speed as follows:

$$a_{idm}(s, v, \Delta v) = a\left[1 - \left(\frac{v}{v_0}\right)^4 - \left(\frac{s^*(v, \Delta v)}{s}\right)^2\right] \quad (1)$$

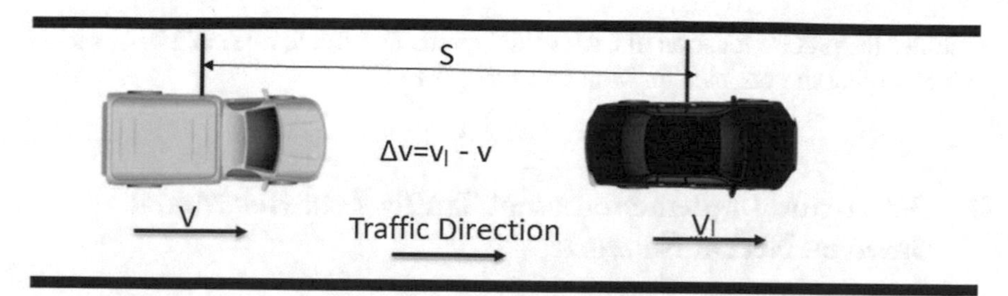

Fig. 1 Diagram of IDM (s is the distance between two consecutive cars while v_l, v, $\Delta v\Delta v$ are the velocities of the former car and the rear car, and the difference between them, respectively)

$$s^*(v, \Delta v) = s_0 + vT + \frac{v\Delta v}{2\sqrt{ab}} \tag{2}$$

In Eqs. 1 and 2, s_0 represents the allowable minimum distance between two vehicle, a and b are parameters decided by manual adjustment and T represents the reaction time of the driver. As seen in Eqs. 1 and 2, the acceleration can be divided into two parts: the first part $a_{acc} = a\left[1 - \left(\frac{v}{v_0}\right)^4\right] a_{acc} = a\left[1 - \left(\frac{v}{v_0}\right)^4\right]$ represents the smooth acceleration toward the target speed and the second part $a_{dec} = -a\left(\frac{s^*}{s}\right)^2 a_{dec} = -a\left(\frac{s^*}{s}\right)^2$ represents the deceleration strategy based on the current distance s and the target distance s^*. The target distance s^* is a variable relating to the current vehicle speed v and the difference with the preceding vehicle speed $\Delta v\Delta v$. While accelerating towards a target speed on a smooth road, the acceleration gradually decreases from the initial maximum to zero. The IDM shows a steady, collision-free dynamic model under a smart deceleration strategy. Figure 1 is a schematic diagram of the IDM.

2.2 *United States Highway Traffic Dataset*

Next Generation Simulation (NGSIM) is an organization focused on traffic data acquisition and analysis. The core objective of the organization is to develop a set of open-source traffic behavior algorithms based on microscopic models that researchers from around the world have free access to. The dataset used in this chapter is from the United States (U.S.) Highway 101 (Xu 2006).

U.S. Highway 101 is located in Los Angeles, California, and is also known as the Hollywood Expressway. NGSIM collected traffic data from one particular section of the highway on July 15, 2005. The studied area is 640 meters long and contains five main roads. Researchers placed eight digital cameras on a 36-storey building on both sides of the road to capture images of vehicles, then used a self-developed software called 'NG-VIDEO' to extract the car's orbit data from the video. The data includes the precise location of each vehicle, reaching frame rates of 10 per second and recording the vehicles' information every 0.1 s.

3 Design and Implementation of Traffic Behavior Model Based on Neural Network

3.1 Overview

The popularization of vehicles in people's daily life has shifted more attention to traffic behavior models and traffic visualization technologies. In the field of computer graphics, research on transportation basically focuses on two topics: traditional traffic simulation and traffic reconstruction. The way traffic flows evolve given a specific traffic network, a traffic behavior model, and the initial status of the vehicles is the core problem of traffic simulation. It is the micro-traffic model that this chapter focuses on. The micro-model concentrates on how a car is affected by the other vehicles around it. Traditional traffic models generally use mathematical equations to calculate current vehicle accelerations based on the state of the surrounding vehicles, with these equations based on certain assumptions. Meanwhile, the parameters involved are also determined from human experience. The traffic flows simulated by traditional models do not effectively reflect the characteristics of the complex and ever-changing real traffic flow, for models based on mathematical equations do not utilize real traffic orbit data. When assumptions are not sufficiently accurate, simulated traffic flows are vastly different from actual traffic flows. Therefore, traditional traffic flow simulation methods suffer from the following deficiencies:

(1) Based on the one-time initial data input;
(2) All cars are virtual cars and their orbit data are completely calculated by mathematical models;
(3) The simulated traffic flow cannot adequately reflect the complex and ever-changing real traffic flow.

Real traffic flows are very complex situations in which drivers' driving behaviors are affected by many changing factors. In most cases, both the surrounding vehicles as well as road conditions are constantly changing. Drivers may make different decisions, depending on dynamic traffic conditions, resulting in very drastic changes in drivers' driving behaviors. The parameters that govern driver behavior are also constantly changing.

Therefore, in order to simulate traffic flows more realistically, we aimed to learn from real vehicle traffic trajectories and replace the traditional human-defined mathematical models of vehicle behavior. Our model is based on neural networks developed through the mapping between the traffic conditions faced by the vehicle and the decisions made by the vehicle.

The key innovation of this method is developing behavioral models that reflect each vehicle's driving characteristics by using real traffic data for each vehicle. The mapping is achieved through a neural network in which the driver's driving model is implicit.

After studying the traffic behavior of each vehicle model, the interaction between virtual and reality is introduced into the traffic model through the insertion of a real car and a virtual car. The real car's behavior trajectory is set to be immutable. The interaction between virtual car and real car is controlled by the model that was developed from the neural network.

Traffic simulation is an important branch and research hotspot of computer graphics. From the simulated traffic flow, we can evaluate the validity of the traffic model used. In traditional traffic models like IDM, the driving behavior of a car is mainly affected by the leading car while in actual driving, the behavior of a car may not only be affected by the car in front but also the car behind. Therefore, in order to better characterize a driver's driving behavior, this chapter introduces a three-car following model in which the behavior of a car is affected by the front and rear cars. In addition, using real-time trajectories in the traffic flow is also different from the traditional method where researchers usually input only the vehicle's initial settings like its position, speed, and acceleration, letting the evolution of traffic flow be calculated by the simulation model. It is noteworthy that traffic flow simulated this way is relatively gentle and thus reflects real traffic conditions to a lesser extent. The precondition of our simulation system is that the real traffic trajectory is unchangeable, so that researchers can mainly focus on interaction between real cars and virtual cars.

3.2 *Data Collection and Pretreatment*

3.2.1 Data Collection

Table 3.1 is the format of the original traffic data set. Each row of the original dataset records the state of a car for a moment, for a total of 18 rows.

The data used in this study was captured between 8:05 AM and 8:20 AM Pacific Daylight Time (PDT) on July 15, 2005. The videos were recorded at a rate of 10 frames per second, capturing textual vehicle information.

To obtain the necessary datasets of three consecutive cars each, vehicles' driving data were extracted from the videos. This data included factors that could influence the drivers' decision making, including adjacent vehicles' speed differences, distances, and accelerations. In order to enhance the quality of extracted data, the following requirements had to be met:

(1) There are cars in the same lane as and in front of and behind the target car;
(2) The distance between the car and the target car is less than 120 m;
(3) The front and rear cars are travelling at speeds under 20 km/h;

Table 3.1 Original traffic data formation

Variable	Description	Type of data
Car_ID (A)	Vehicle's identification number	int
Frame_ID (B)	Vehicle's frame number	int
TotalFrame (C)	The total number of frames that the car appears in the video	int
GlobalTime (D)	The total time that the car appears in the video	int
Local_x (E)	Horizontal distance between the car and the left edge of the lane	double
Local_y (F)	Vertical distance between the car and the start point of the lane	double
Global_x (G)	Horizontal global coordinates of the car center, based on CA State Plane III	double
Global_y (H)	Vertical global coordinates of the car center, based on CA State Plane III	double
Length (I)	The length of car, in feet	double
Width (J)	The width of the car, in feet	double
TypeofCar (K)	'1' represents motorbike, '2' represents common car, '3' represents truck	int
Speed (L)	The speed of car, in feet per second	double
acc (M)	The acceleration of car, in feet per second squared	double
lane (N)	There five lanes in total	int
Front_ID (O)	The current car's proceeding car ID, '0' means there is no car in front	int
Rear_ID (P)	The current car's rear car ID, '0' means there is no car	int
SpaceInterval (Q)	The distance between the current car center and the preceding car center	double
TimeInterval (R)	The time it takes for the current car to travel to the front car center at current speed	double

(4) The distance and time between the target car and its front and rear vehicles is less than 61 m and cannot be less than 30 s, respectively.

3.3 Design of Neural Network

3.3.1 Input and Output of the Network

Drivers' driving objectives vary with the surrounding traffic conditions. For example, in crowded traffic conditions, the driver's goal is usually to follow the preceding vehicle. On the other hand, when the car ahead suddenly decelerates, the driver's goal is to avoid collision.

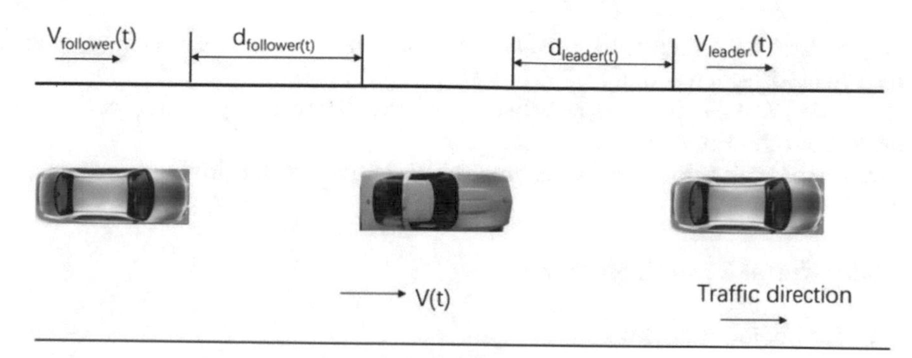

Fig. 2 Diagram of input variables for the neural network

This chapter argues that traffic behavior model can be solved by mapping from traffic status to driver behavior. The driver's driving behavior is affected by traffic conditions which can be represented by a set of variables that characterize the dynamics of the vehicle and its surroundings. In the following model, traffic states are generally represented by the distance to the preceding vehicle, the relative speed, and the relative acceleration.

The input of the model should be the most influential factor for the driver's driving behavior. In reality, a driver's behavior can be affected by many factors such as the surrounding vehicles, the driver's fatigue level, the road conditions, the driver's preference, the type of car, and the presence of companions. Some factors are observable while others are not. We take advantage of the characteristics of the dataset to utilize the factors that affect drivers' driving behavior most to maximize accuracy.

In order to better represent the traffic conditions, we use a three-car following model. The output of the network is the decision that a driver makes at a specific time, expressed in terms of acceleration, speed or distance. Figure 2 is a schematic diagram of the input variables of the neural network used. Due to a certain degree of noise in the data source, the variable to be chosen as the output of the model should be based on the principle of minimum error (this article uses vehicle speed), while the function to be learned by the neural network is shown in Fig. 2:

$$v_{(t)} = f(\Delta v_{follower}, \Delta s_{follower}, a_{follower}, \Delta v_{leader}, \Delta s_{leader}, a_{leader},) \quad (3)$$

$v(t)$: The current vehicle driving direction at time t, also the output of the neural network;
Δv_follower: The difference in speed between the current car and its rear car in the driving direction at time t;
Δs_follower: The relative distance between the current car and its rear car in the driving direction at time t;
*a_followe*r: The acceleration of the vehicle after time t in the driving direction;

Δv_leader: The difference in speed between the current car and its preceding car in the driving direction at time *t*;
Δs_leader: The relative distance between the current car and its preceding car in the driving direction at time *t*;
a_leader: Acceleration of the preceding vehicle at time *t* in the driving direction.

3.3.2 Neural Network Structure

(1) The Choice of Network Layer

The choice of network layer is still of no mature theoretical guidance. It is generally thought that as the depth of the network layer increases, the model's strength increases correspondingly, leading to smaller and fewer errors. However, the model becomes prone to over-fitting and needs a longer training time. The optimum number of network layers must be determined through a test in which the number of network layers is increased from the initial three-tier network (with a single hidden layer) to observe resultant network performance improvements. In theory, the same effect can be achieved by increasing the number of nodes in a single hidden layer while simultaneously increasing the number of layers in the network.

In the experiment described in this chapter, we started with a single hidden layer then gradually increased the number of hidden layers while reducing the number of nodes in the hidden layers. It was found that increasing the number of hidden layers did not greatly improve the network performance, in addition to prolonging the training time. In conclusion, the mapping described in this chapter is relatively simple with only a single hidden layer.

(2) The Choice of Hidden Layer Nodes

In the backward propagation neural network, a high number of hidden layer nodes leads to over-fitting which greatly influences the neural network's performance. Therefore, the choice of hidden layer node quantity is particularly crucial. Regardless of whether the number of hidden nodes is large, there is currently no robust set of theories that accurately calculates the number of hidden. Many researchers have put forward various calculation methods which have unfortunately shown poor fit when tested on training sets. A single data set can be used in dozens or even hundreds of ways via different calculation methods.

The number of nodes in a hidden layer is affected by many factors: the size of the data set, the complexity of the problem to be solved, the selection of the activation function, and the number of nodes in the input and output layers.

In order to improve the neural network model's performance and generalization ability and to avoid over-fitting, the number of hidden nodes used was minimized as much as possible without affecting network performance. The following conditions needed to be met while calculating the number of hidden layer nodes:

(1) The number of training samples should be one unit greater than the number of hidden nodes. Otherwise, the characteristics of the training sample do not work and the network capability will gradually decrease to zero, regardless of the sample. However, networks trained in this way have no generalization ability. Similarly, the number of nodes in the input layer should also be one unit less than the number of samples.
(2) The number of network weight parameters should be less than the number of training data, usually by half to a tenth. Otherwise, a stable network model will not be achieved unless training samples are divided into several smaller samples for the use of the "rotation training" method.

In short, if there are too many nodes in the hidden layer of the network, the network model becomes overly complex and a longer training time results. Even if the error of the network is reduced, the overfitting phenomenon will occur and the model's generalization ability will be weakened. If there are too few hidden nodes in the network, the network cannot be trained or the performance is poor. Therefore, in the actual network training process, node expansion and deletion methods are necessary to determine the number of hidden layer nodes, in order to avoid overfitting.

In the actual training process, as the dataset size of each vehicle was generally around 400 points, the number of input nodes was 6, and there was one output, the number of hidden layer nodes was controlled within [5, 50]. In training, by starting from 5 and gradually increasing the number of hidden nodes to observe changes in the error function, we found that more than 10 hidden nodes produced smaller network performance increases and extended training times, so the number of hidden nodes was finalized as 10.

(3) Determination of Loss Function

In the network, a loss function was defined as the difference between the actual network output and its expected output, with network optimization being performed in response to the loss function. In terms of classification, a classical loss function is a cross-entropy function. Regression problems involve making specific numerical predictions, such as price forecasting, sales forecasting and so on. The prediction of vehicle behavior is also a regression problem. These questions need to predict not a predefined category but a real number. The output of this neural network is the predicted acceleration. For the network of regression problems, the most commonly used loss function is the mean square error (MSE), defined as shown in Eq. 4:

$$MSE(y, y') = \frac{\sum_{i=1}^{n} (y_i - y'_i)^2}{n} \tag{4}$$

where y_i is the correct answer for the i-th data given by the neural network and y' is the expectation value of the answer.

4 Model Simulation

After developing a behavior model for each vehicle using the neural network, the current traffic status of the vehicle is input to calculate the speed that the vehicle should take at this point in time. Subsequently, the target vehicle's ideal acceleration and distances from its preceding and following vehicles are calculated. During the simulation, the position status of each car is calculated using Eq. 5:

$$\mathrm{p}(\mathrm{t}+\Delta t) = p(\mathrm{t}) + v(t)\Delta t \tag{5}$$

where $p(t)$ represents the position of the vehicle at time t and v(t) equals the velocity of the car at time t. In the simulation process, Δt is set to 0.1 s.

Vehicle driving characteristics developed from real traffic trajectories can be easily applied to large scale traffic flow simulation systems that are based on micro-models. At present, there are ongoing research efforts to find suitable driving strategies for individual vehicles with respect to their minimum acceleration and maximum safety distance, among other outputs. However, it must be noted that some of the models are overly ideal and thus do not reflect real traffic environments.

The goal of this approach is to reflect more realistic traffic scenarios than to find the best trajectory for each car. The real traffic data of each car was used by the neural network to find the car's driving characteristics. With these individual vehicle behavior models, the neural network method is able to simulate the traffic situation at any time and place. In general, traffic reconstruction is a scene reconstruction of the sample video. However, to a deeper degree, the sample-based traffic simulation can be viewed as a scene extension in space and time.

5 Model Verification

Model verification and evaluation is an important step in the model construction process. Model evaluations are conducted with qualitative and quantitative methods. In this study, the strength of the neural network model was qualitatively examined through the traffic trace curve and quantitatively through mathematical statistics. The combination of these two methods produced a comprehensive evaluation, thus allowing for the comparison of the neural network model with other traffic models to determine its advantages.

The training data was provided by NGSIM and comes from Highway 101 near University Park in Los Angeles, California, between 8:05 AM and 8:20 AM PDT on July 15, 2005. The traffic data for each car was recorded by NGSIM-VIDEO at a rate of 10 frames per second. Due to the different speed limits in different regions, local road conditions and driving habits also affect the driving behavior of vehicles.

Our method aims to increase the authenticity of simulated traffic flows. Changes in acceleration could be a good indicator of the complexity of traffic conditions.

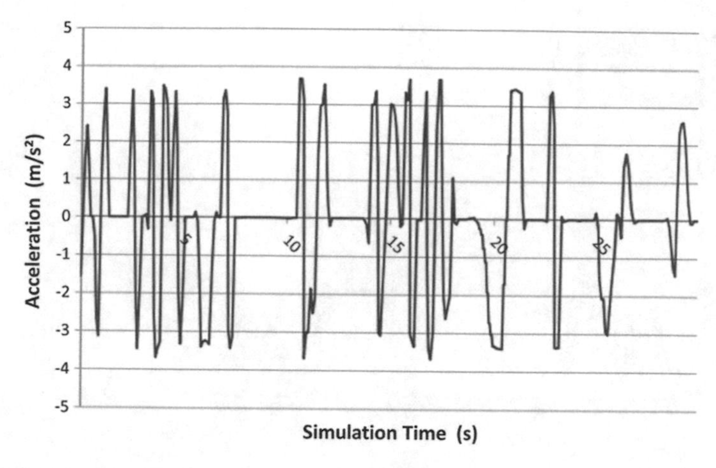

Fig. 3 Vehicle acceleration in reality

Acceleration also reflects the driving habits and characteristics of different drivers. The data in Fig. 3 shows a real car's acceleration over time, from which it can be concluded that the traffic conditions each vehicle faces in reality change rapidly, and that the acceleration of a vehicle is unpredictable. The traditional method of modeling traffic flows based on mathematical equations is inadequate for this task.

Driving models were developed from 50 vehicles' real traffic trajectories. The first 70% of each car's data was used as a training set, and the remaining 30% as a test set. Two vehicles (Cars 21 and 23) were randomly selected for use in a comparison of the neural network method with another method. Developed and studied by Bi et al., this second method (hereon called the Bi method) is a different data-driven method of modeling traffic behavior (2016) that predicts the status of the target vehicle based on the state of the vehicle in front of the vehicle. In the Bi method, one model is applied to all vehicles while the proposed neural network method takes into account different vehicles' personalized driving habits, generating a behavior model for each vehicle that best reflects its characteristics.

The two methods were compared using their predictions of the target car's speed (see Figs. 4a and 5a and the distances between the target car and its preceding and following cars (see Figs. 4b, c and 5b, c). From Figs. 4a and 5a, it can be seen that the Bi method does not accurately reflect the driving characteristics of a single vehicle, likely due to its use of the data from all vehicles as a unified training set for model development which produced a state of average vehicle behavior. The low accuracy of the Bi method in predicting vehicle speed naturally led to the low accuracy of its predictions of vehicle distances as well, as seen in Figs. 4b, c and 5b, c.

Figure 5 shows the comparison result of Car 21. It can be seen from the speed comparison that during the whole simulation, the results of the Bi et al. method always start higher than the real value and lower than the value predicted by our system. Meanwhile, the simulation based on neural network in our proposed method

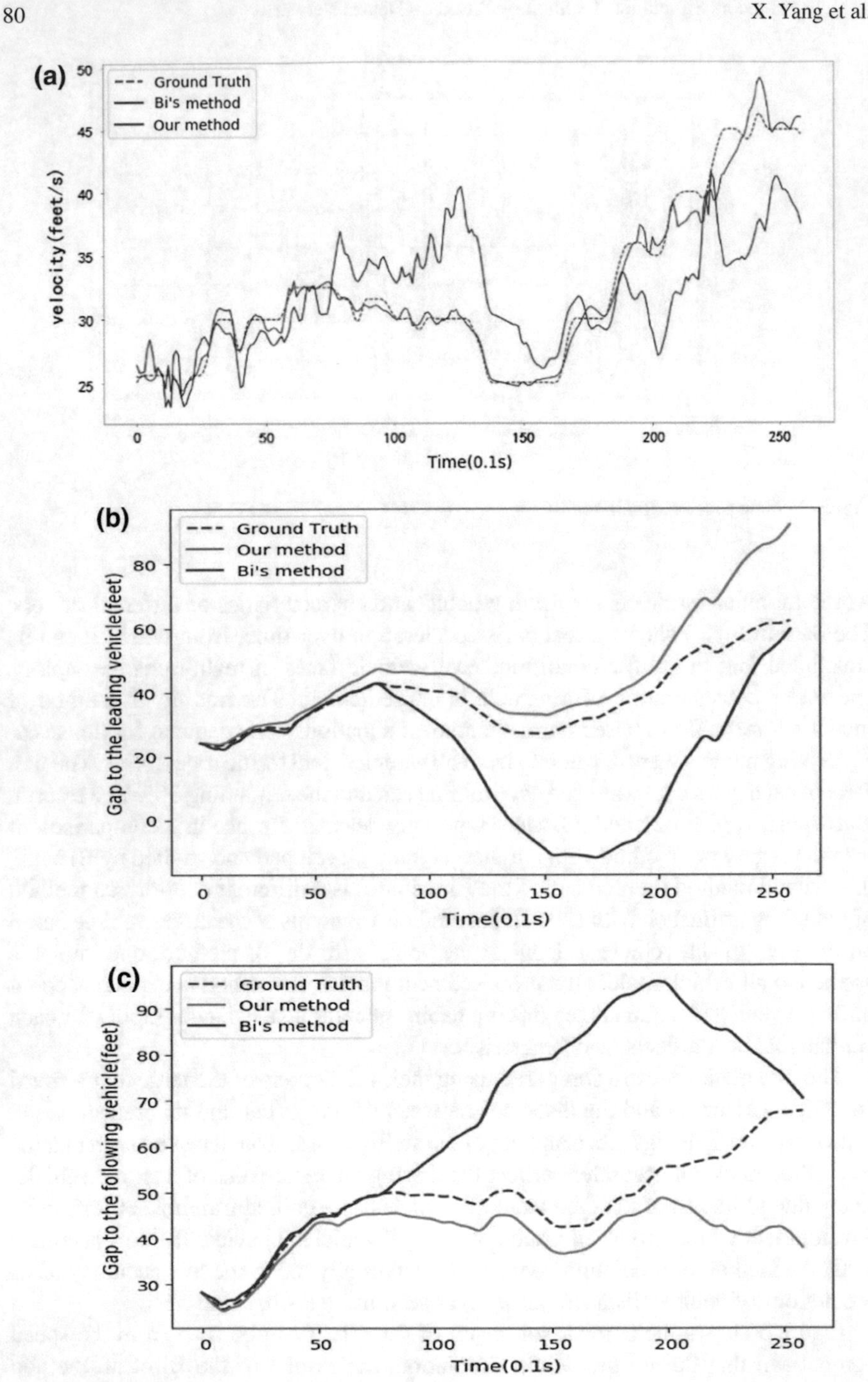

Fig. 4 **a** Speed of Car 23. **b** Distance between Car 23 and its preceding car. **c** Distance between Car 23 and its following car

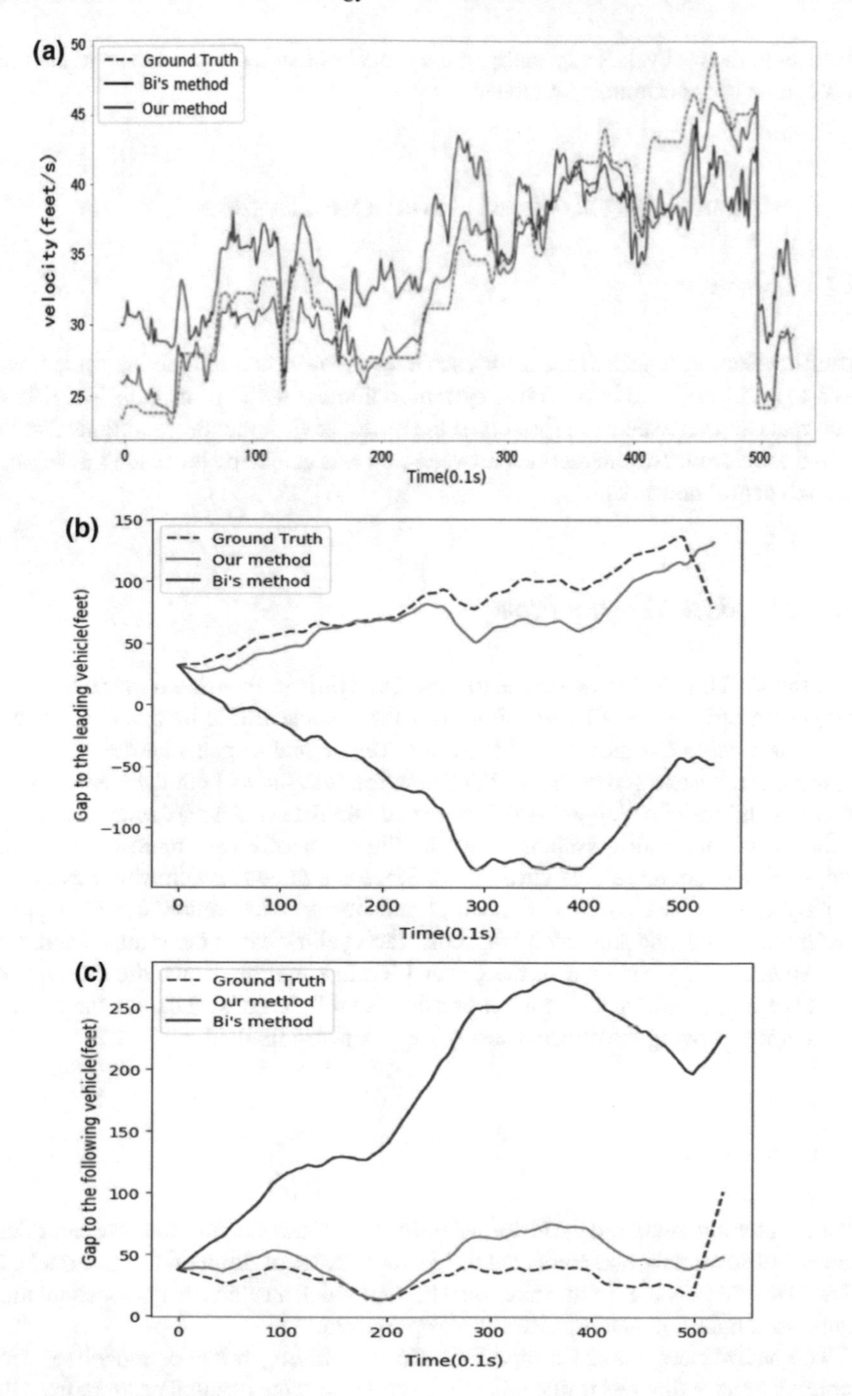

Fig. 5 **a** Speed of Car 21. **b** Distance between Car 21 and its preceding car. **c** Distance between Car 21 and its following car

is closer to the real value, especially in the process of sudden speed drop in the final part can be very accurately simulated.

6 Design of the Real-Virtual Interaction System

6.1 Overview

After developing a traffic model for each car, a real-virtual interaction system was used to insert real cars into a traffic system to interact with virtual cars. The data of each real car was fixed and represented the traffic environmental conditions that the virtual cars faced. The interactions between cars was guided by the models developed through neural networks.

6.2 Simulated Traffic Flow

We used the Unity3D game engine to generate a three-dimensional traffic flow animation. In order to avoid animation lags, the vehicle traffic data was calculated in advance using the model, then input into the engine to generate the animation. Figure 6 is a system screenshot of the animation that shows both the Ground Truth situation and the situation predicted by virtual simulation at a particular instant.

In the virtual reality system shown in Fig. 6, purple cars represent real cars while yellow cars represents virtual cars. Since the three-car following model was employed, there is a purple car ahead of and behind each yellow car. Comparing the Ground Truth and simulated snapshots, each yellow car in the virtual simulation corresponds to a purple car in the Ground Truth situation. Thus, the behavior of virtual cars in the simulated result is consistent with that of real cars in the Ground Truth, again proving the effectiveness of our proposed method.

7 Conclusion

This chapter first reviewed the history of traffic model development and the classification of traffic models, and compared the characteristics of different traffic models. In view of the deficiencies of the traditional traffic model, a vehicle behavior simulation framework based on neural networks was proposed.

The neural network-based model develops a driving behavior model for each vehicle by extracting real traffic trajectories and designing a neural network from the real data of each vehicle to generate a specific driving model. A real-virtual interaction system was then introduced to use the model developed to guide the interaction

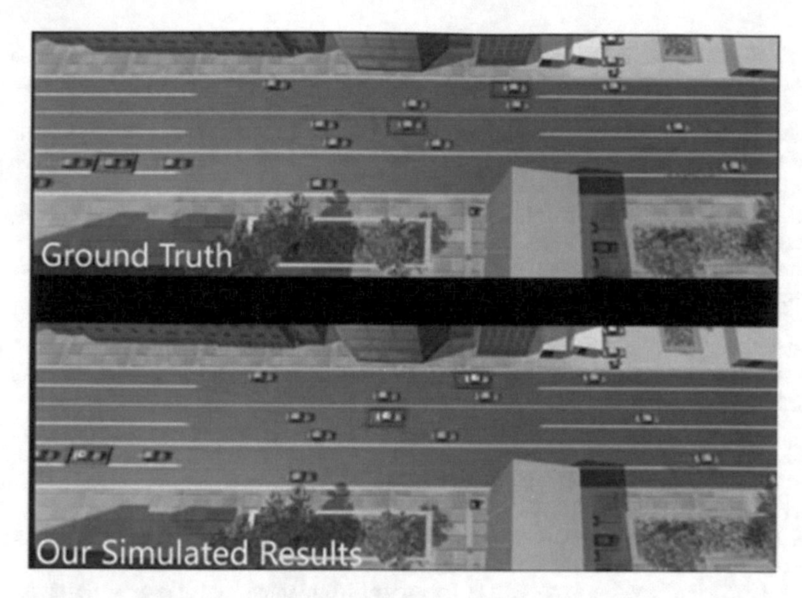

Fig. 6 Real-virtual interaction system

between a real car and a virtual car. The proposed model has been qualitatively and quantitatively analyzed and evaluated in the experiments.

References

Bando, M., Hasebe, K., Nakayama, A., et al. (1995). Dynamical model of traffic congestion and numerical simulation. *Physical Review E, 51*(2), 1035.

Bi, H., Mao, T., Wang, Z., et al. (2016). *Proceedings of the ACM SIGGRAPH/Eurographics symposium on computer animation*. Eurographics Association, pp. 149–158.

Daganzo, C. F. (1995). The cell transmission model, part II: Network traffic. *Transportation Research Part B: Methodological, 29*(2), 79–93.

Gazis, D. C., Herman, R., & Rothery, R. W. (1961). Nonlinear follow-the-leader models of traffic flow. *Operations Research, 9*(4), 545–567.

Gerlough, D. L. (1955). *Simulation of freeway traffic on a general-purpose discrete variable computer*.

Hidas, P. (2005). Modelling vehicle interactions in microscopic simulation of merging and weaving. *Transportation Research Part C: Emerging Technologies, 13*(1), 37–62.

Ju, E., Choi, M. G., Park, M., et al. (2010). Morphable crowds. *ACM Transactions on Graphics (TOG) ACM, 29*(6), 140.

Kesting, A., Treiber, M., & Helbing, D. (1928). Enhanced intelligent driver model to access the impact of driving strategies on traffic capacity. *Philosophical Transactions of the Royal Society of London A: Mathematical, Physical and Engineering Sciences, 2010*(368), 4585–4605.

Kesting, A., Treiber, M., & Helbing, D. (2007). General lane-changing model MOBIL for car-following models. *Transportation Research Record: Journal of the Transportation Research Board, 1999,* 86–94.

Lebacque, J. P. (1997) *A finite acceleration scheme for first order macroscopic traffic flow models.*

Lee, K. H., Choi, M. G., & Hong, Q., et al. (2007). Group behavior from video: a data-driven approach to crowd simulation. In *Proceedings of the 2007 ACM SIGGRAPH/Eurographics symposium on Computer animation. Eurographics Association* (pp. 109–118).

Lighthill, M. J., & Whitham, G. B. (1955). On kinematic waves. II. A theory of traffic flow on long crowded roads. In: *Proceedings of the Royal Society of London A: Mathematical, Physical and Engineering Sciences. The Royal Society* (Vol. 229, Issue No. 1178, pp. 317–345).

Nagel, K., & Schreckenberg, M. (1992). A cellular automaton model for freeway traffic. *Journal de Physique I, 2*(12), 2221–2229.

Nelson, P., Bui, D. D., & Sopasakis, A. (1997). *A novel traffic stream model deriving from a bimodal kinetic equilibrium*.

Newell, G. F. (1993). A simplified theory of kinematic waves in highway traffic, part I: General theory. *Transportation Research Part B: Methodological, 27*(4), 281–287.

Next Generation Simulation, US Highway 101 Dataset, http://www.fhwa.dot.gov/publications/research/operations/07030/.

Pipes, L. A. (1953). An operational analysis of traffic dynamics. *Journal of Applied Physics, 24*(3), 274–281.

Prigogine, I., & Andrews, F. C. (1960). A Boltzmann-like approach for traffic flow. *Operations Research, 8*(6), 789–797.

Qiao, Jin. (2008). *Research on calibration and verification of car-following model parameters*. Shanghai: Shanghai Jiao Tong University.

Sewall, J., Wilkie, D., & Lin, M. C. (2011). Interactive hybrid simulation of large-scale traffic. *ACM Transactions on Graphics (TOG) ACM, 30*(6), 135.

Sewall, J., Wilkie, D., & Merrell, P., et al. (2010) Continuum traffic simulation (Vol. 29, Issue No. (2), pp. 439–448). Computer Graphics Forum. Blackwell Publishing Ltd.

Shen, J., & Jin, X. (2012). Detailed traffic animation for urban road networks. *Graphical Models, 74*(5), 265–282.

Treiber, M., & Helbing, D. (2001). Microsimulations of freeway traffic including control measures. *at-Automatisierungstechnik Methoden und Anwendungen der Steuerungs-, Regelungs-und Informationstechnik, 49*(11/2001), 478.

Xu, X. (2006). *Study on car-following model based on artificial neural network*. Beijing: Beijing University of Technology.

Zhou, Y. (2007). *Several key technologies in intelligent vehicles*. Shanghai: Shanghai Jiao Tong University.

BlockTower: A Multi-player Cross-Platform Competitive Social Game

Ruichen Zheng, Hai-Ning Liang, Ruiqi Xie, Feiyu Lu, Yuwei Shi, Wenge Xu and Konstantinos Papangelis

Abstract This research aims to explore the usability and gameplay experiences of a two-player game across three platforms. The game is based on a variation of the board game Jenga—a game played in social, casual settings. Our game requires spatial, tactical, and mental reasoning from players to win and consists of a tower of blocks piled on top of each other. Players take turns to remove one block at a time from the tower then place it on the top of the tower; the game ends when at least one block of the tower falls to the ground—the losing player is the one whose turn caused this fall. The game can be played by two players simultaneously through the use of virtual reality head mounted displays (HMD) like the Oculus RIFT or HTC Vive, two separate mobile tablet devices, or a shared television (TV) display. The first two platforms provide players with their own independent views of the game, whereas the third platform allows the players to use one single screen. This research is an attempt to better understand how virtual reality and mobile technologies can support multiplayer gaming experiences when gamers are engaged in a competitive casual game.

Keywords Virtual reality · Head-mounted displays · Mobile tablets
Multiuser gameplay · Social gaming

1 Introduction

There has been growing appreciation for the use of virtual reality head-mounted displays (HMD) such as the Oculus RIFT (Oculus Rift | Oculus 2016) and the HTC Vive (Vive | Discover Virtual Reality Beyond Imagination 2016) to enhance and

---

R. Zheng · H.-N. Liang (✉) · R. Xie · Y. Shi · W. Xu · K. Papangelis
Department of Computer Science and Software Engineering, Xi'an Jiaotong-Liverpool University, Suzhou, China
e-mail: haining.liang@xjtlu.edu.cn

F. Lu
Department of Computer Science, Virginia Tech, Blacksburg, VA, USA

Y. Cai et al. (eds.), *VR, Simulations and Serious Games for Education*, Gaming Media and Social Effects, https://doi.org/10.1007/978-981-13-2844-2_8

add variety to users' gaming experiences (Tan et al. 2015). Due to the technology's rapid development, the types and variety of games that are compatible with HMD are currently limited. In addition, the types of gaming experiences that VR technologies can lead to are still relatively unknown. In this research, we explored the affordances and gameplay experiences of a game designed for two players using VR HMD. We also aimed to investigate differences between the VR experiences and those produced by mobile tablet use and TV use (Lindt et al. 2005), which represent two of the most popular settings for multiplayer casual gaming (Kuittinen et al. 2007). While the large TV setting remains the most common way of playing console games, gaming on mobile tablets has become more mainstream, as exemplified by the introduction of devices like the Nintendo Switch—a console that is also a gaming tablet.

This project explored the three gaming platforms (i.e., VR, mobile tablet, and TV) and their corresponding gameplay settings when two gamers play a competitive casual game. The chosen cross-platform game would have a separate user interface depending on the interaction affordances of each platform (Lindt et al. 2006). To this end, we have developed a variation of the Jenga board game which requires spatial, tactical, and mental reasoning from players to win the game. The original Jenga game consists of a collection of rectangular blocks piled on top of each other, creating a tower of three-block layers (see Fig. 1). Players take turns to extract one block at a time while ensuring that the tower does not become so unstable that blocks fall off the tower. As in the original game, the players need to put the blocks that have been taken out back onto the top of the tower and can place them at any angle that is physically possible so that it is challenging for later blocks to be placed on top. As our game utilizes physics laws, players can use their physics knowledge to place blocks in sophisticated and difficult positions on top of the tower so as to make it challenging for the other player to position blocks during their turn. The game ends when a block falls to the ground. The game allows gameplay via HMD, mobile tablets, or a large TV display. Physical distance between the players is not an impediment as the first two settings allow for remote gameplay.

In this chapter, we introduce our game, BlockTower, and describe how players can play it on each of the three platforms. We also report the results of a study with the game.

2 BlockTower

BlockTower is a turn-based game for two players. Our implementation of the game consists of 18 blocks placed in groups of three to make up each level of the tower (thus, gameplay begins with a tower containing six levels of three blocks each). When it is a player's turn, they need to select a block and take it out of the tower, then place it on the top of the tower. The system calculates if any block in the tower would fall or whether the whole structure is stable enough that no block will fall onto the ground. If a block falls on the ground, the player loses the game. If the tower is stable, the other player takes their turn. The game obeys all laws of physics, besides

Fig. 1 The Jenga game being played by two players. Each player takes turns to remove a block from the tower and place it back on the top of the tower while making sure the tower remains stable. The game ends when the tower or any piece falls to the ground

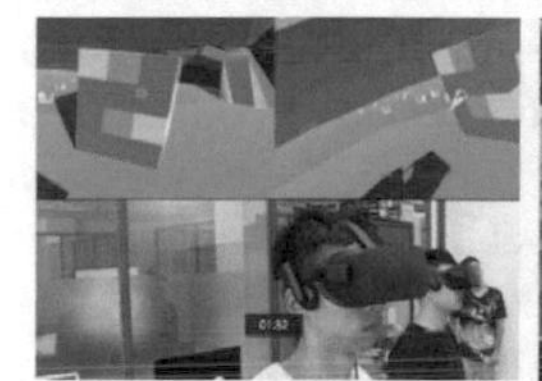
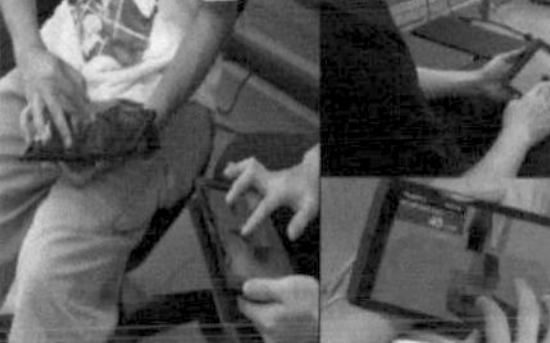

Fig. 2 The three versions of the BlockTower: (Left) Players wearing VR HMD; (Middle) Players using a mobile tablet each; and (Right) Players sitting in front of a large display

allowing the selected block to float in the air while the player is deciding on where to put it, and can be played with the players being physically next to each other or connected via remote networks. Each of the three technology platforms supports a different type of gaming experience. The VR headsets let the player be immersed in the game. The players can change their views of the game by moving their heads up or down, left or right, and closer or further. When players play using a mobile tablet, they need to use the touch capabilities of the device's display to interact with the game elements. In contrast with the VR goggles and tablets where each player has an independent view of the game, the large display platform provides a shared screen on which both players see the same display. However, when players who are using the mobile platform are physically close together, the tablets can be seen as semi-independent as each player can choose to let the other player see their device display. Finally, the controls for the VR and large display platforms are based on an Xbox controller (Fig. 2).

3 Experiment

3.1 Design, Apparatus, and Participants

We conducted a study to examine the usability issues and gameplay experiences of the game on the VR, mobile tablet, and large display platforms. In all three versions, the ray casting technique where the player casts a ray at a point on the two dimensional (2D) display to indicate an object in the three dimensional (3D) scene (Liang and Green 1994; Hinckley et al. 1994) is utilized for block selection. Although the technique is limited in its distant object interaction capabilities (Pierce et al. 1997), it is sufficient for achieving precise results. In our implementation, the ray casting method varied in each version based on the display type. For the VR and large display versions, the ray casting point was indicated by an onscreen cursor as visual feedback. The point cursor was chosen to ensure fast selection times with low moving footprints (Vanacken et al. 2007). In the mobile version, the ray casting point was directly determined by where the player's finger tapped on the screen.

In the VR platform, participants each wore an Oculus RIFT CV1 HMD, which was connected to a computer running an i7 @ 2.7 GHz with 8 GB of memory and a GTX1070 dedicated graphics card. During each player's turn, a ray was cast from the center of the view to generate a cursor on the surface of the first target object (see Fig. 3).

When the cursor became attached to the surface of a target block, clicking button "A" (see Fig. 3, Left) on the game controller would shade that block red to indicate that it had been selected. When a block was selected, it could then be moved forward/backward/left/right (along the X and Z axes) with the use of the left joystick (up/down/left/right) or up/down (along the Y axis) by moving the right joystick (up/down) (see Fig. 4, Right).

The mapping of actions in the game controller control was based on the similarity between the finger movement and the in-game object movements. Note that the selected block would not be affected by gravity in that it would not fall while moved in midair—this was intended for easy manipulation. After deciding where to put the selected block (and hence to finish the turn), the player would need to press the "A" button again to deselect the block, after which the influence of gravity would be restored and the block would ideally fall onto the top of the tower and rest on the blocks below it. If no block continued to move or fell to the ground within a certain period of time, this meant that the tower was stable and that the other player would be able to start their turn (see Fig. 5). No matter which player's turn it was, both players could change their viewing angles and viewing distance by moving their head or walking around.

Each participant in the tablet platform group used an 8 in. NVidia Shield with 1920 × 1200 multi-touch screen resolution, 2 GB of memory, and a 2.2 GHz processor. For this platform, participants played by making gestures on the touch display. One finger tap was used for selecting and deselecting blocks according to the ray casting technique (see Fig. 6).

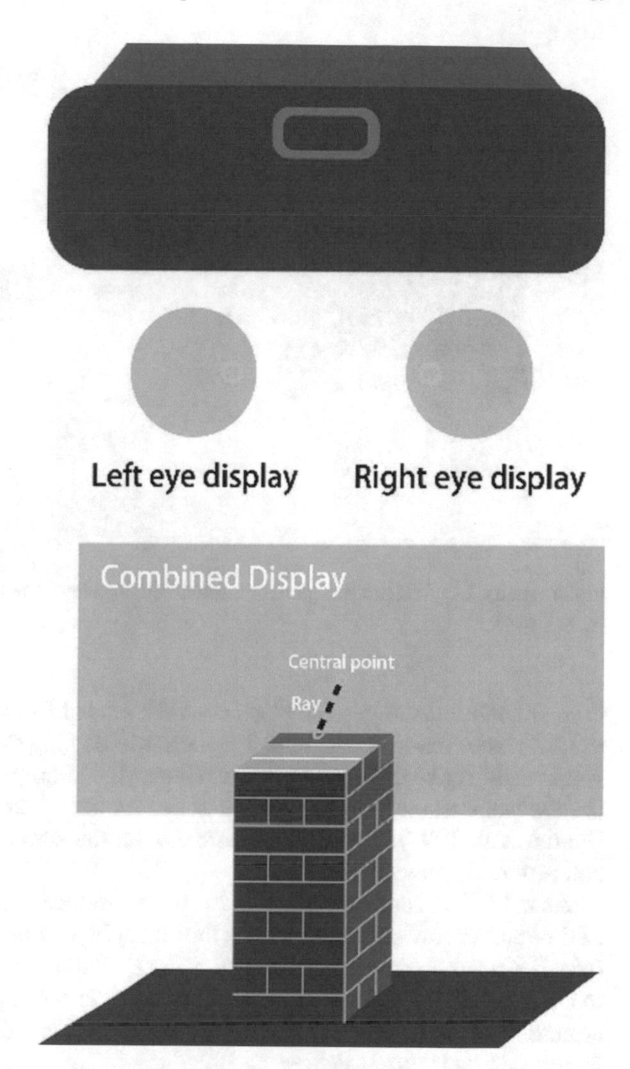

Fig. 3 The ray casting technique implemented in the Oculus and similar VR HMD

The other gestures were designed based on three principles. First, continuous finger position (i.e., dragging) for control mapping was used to maximize performance—this was based on previous research (Zhai 1998). Additionally, other gestures were designed based on their similarities to the transformation of 3D objects in games (Rekimoto 2002). Finally, as recommended in the gestural patterns provided by Remi Brouet for surface interactions (Brouet et al. 2013), the number of touches detected on the screen was used to distinguish ambiguous manipulations and to link associated gestures to actions (see Fig. 7, Right).

In other words, during a player's turn, tapping on the screen would cast a ray from the tapped position and the first block hit by the ray would be selected and freed from gravity. To manipulate the selected block, two types of gestures were possible. By

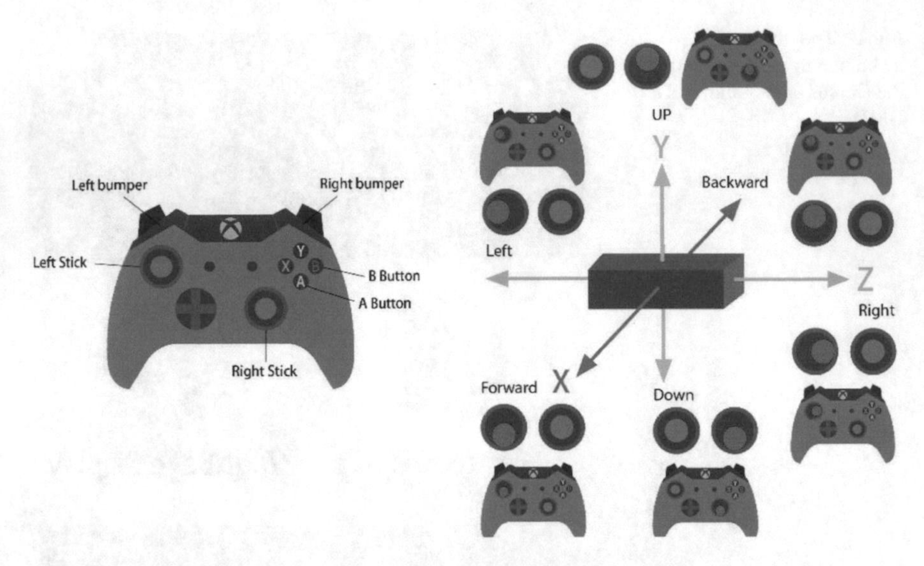

Fig. 4 (Left) A typical modern game controller; (Right) How to move the selected block along the X, Y, Z axes

dragging with one finger, the block could be moved forward/backward/left/right (in the XZ plane) based on the drag direction, while using two fingers to drag up/down would make the block move up or down (along the Y axis) (see Fig. 7). To compensate for the weakness of the ray casting selection technique regarding distant objects (Pierce et al. 1997), tapping anywhere else on the screen would deselect the block and restore its gravity.

As in the VR version, if none of the blocks moved or fell within a certain period of time and the tower became stable, the other player would be able to start their turn (see Fig. 8 for a sample sequence of actions). Although players were still allowed to alter their viewing distance using the two-finger pinch gesture or rotate their view around the tower using the two-finger rotate gesture, their viewing angle became fixed when their turn was over.

Finally, a computer with the same configurations as used for the VR setting was employed in the large display setting. In both the VR and large display settings, participants controlled game elements using Xbox game controllers, as opposed to the tablet version where touch gestures on the display were used as input.

Using the ray casting technique, we placed an onscreen cursor on the shared display. The cursor was controlled by the left joysticks of both players' game controllers. However, only the player whose turn it was would be able to control the cursor; the other player had to wait for their turn to gain the control. The basic controls were a combination of the VR and mobile versions' controls. By clicking the "A" button, a ray would be cast from the cursor position and the first block hit by the ray would be selected and freed from gravity (Fig. 9).

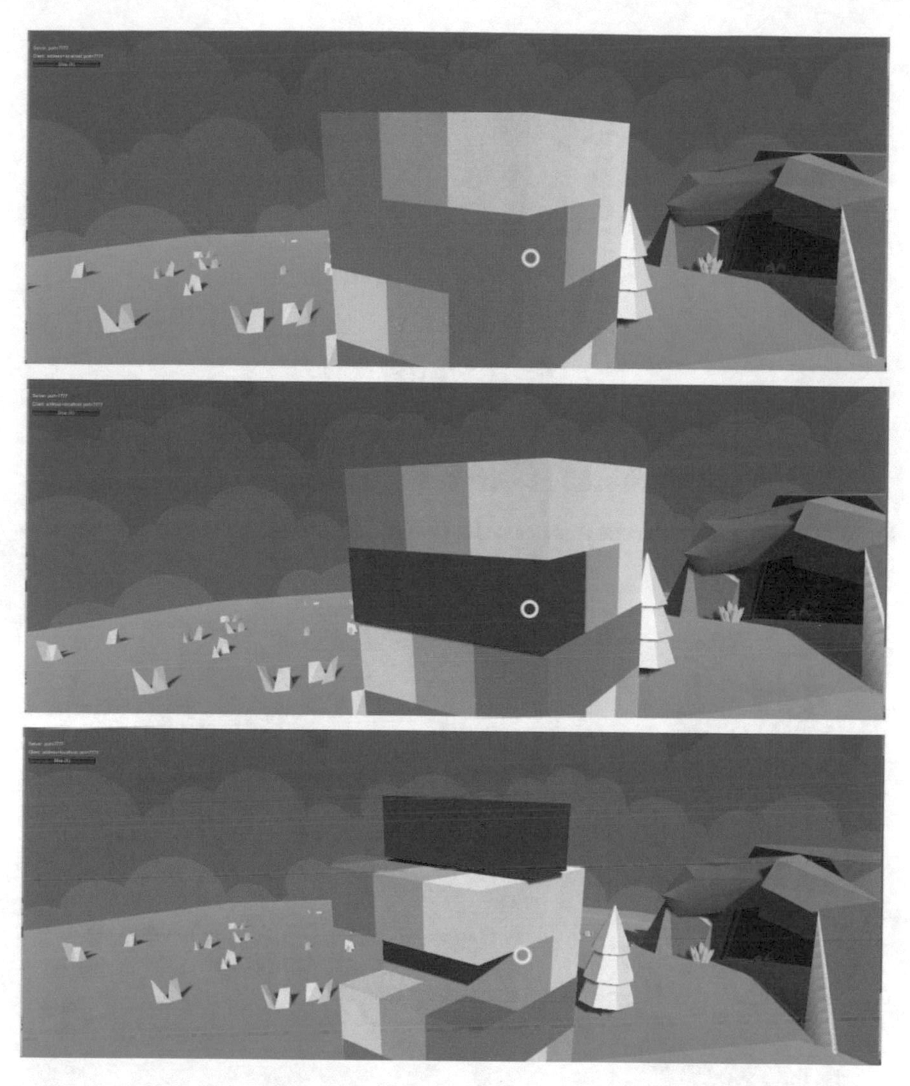

Fig. 5 Screenshots of a sequence of actions: (Top) The cursor represented by the blue circle positioned on a block to be selected; (Middle) The selected block highlighted in red; and (Bottom) The selected block moved and placed on top of the tower

To move a selected block forward/backward/left/right (along the X and Z axes), the player would need to trigger the left joystick (up/down/left/right). On the other hand, by triggering the right joystick (up/down), the player could move the block up/down (see Fig. 10).

Blocks could be deselected and the turn would be ended by clicking the "A" button. If none of the blocks moved or dropped within a certain period of time and the tower remained stable, the other player would be able to start their turn. Regardless of

Fig. 6 The ray casting technique implemented in mobile tablets. Touching a block to enable selection

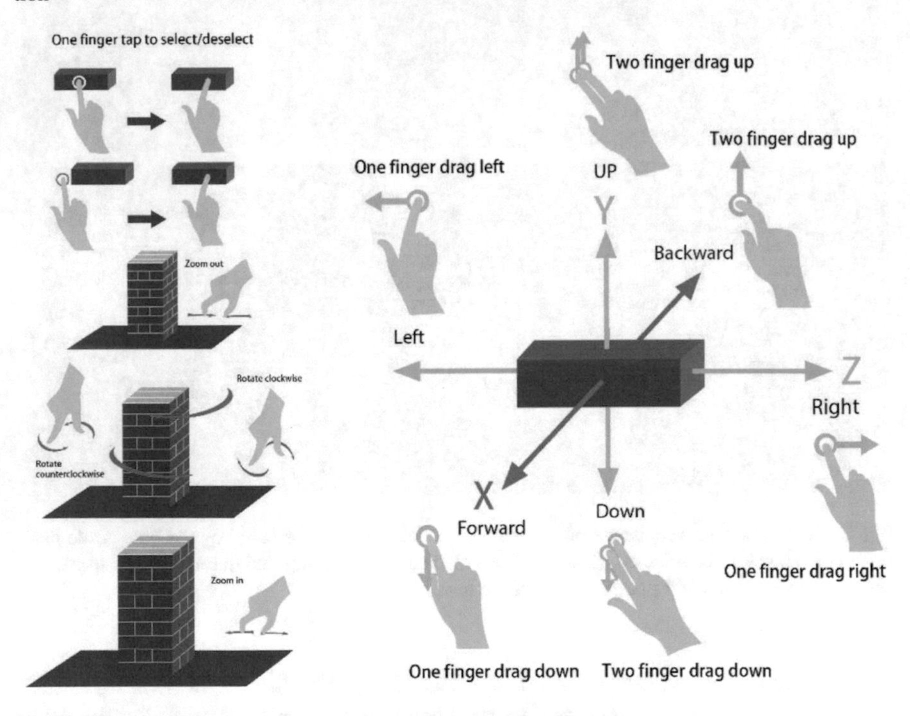

Fig. 7 The gestural set used in our tablet implementation of the game

whose turn it was, both players were only allowed to rotate their view around blocks clockwise/counterclockwise by clicking the "Left bumper/Right bumper" buttons (see Fig. 11).

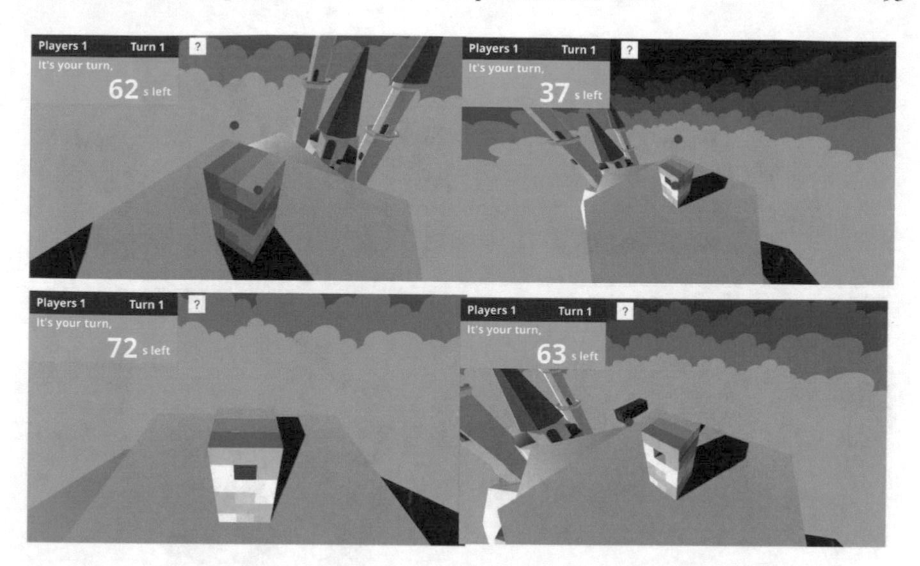

Fig. 8 Screenshots of a sequence of actions in the tablet version: (Top) zooming in and out to find a block to select and move; (Bottom) the selected block being moved with the view of the tower rotated

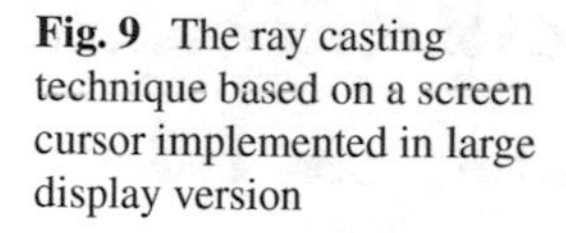

Fig. 9 The ray casting technique based on a screen cursor implemented in large display version

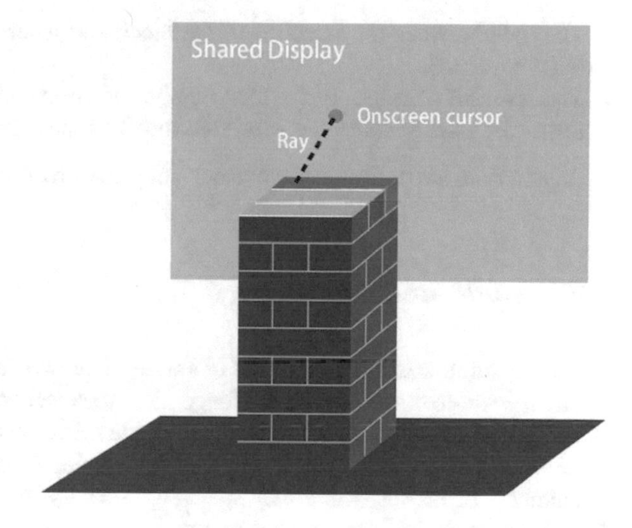

Six pairs of students (all males with an average age of 23) from a local university participated in our experiment; two pairs interacted with each platform. They were given some time to practice and become familiar with the systems, devices, and controllers. Each pair was asked to play the game twice. During the process, we recorded the following:

- The time between the start of a player's turn and when a block was selected (Selection time).

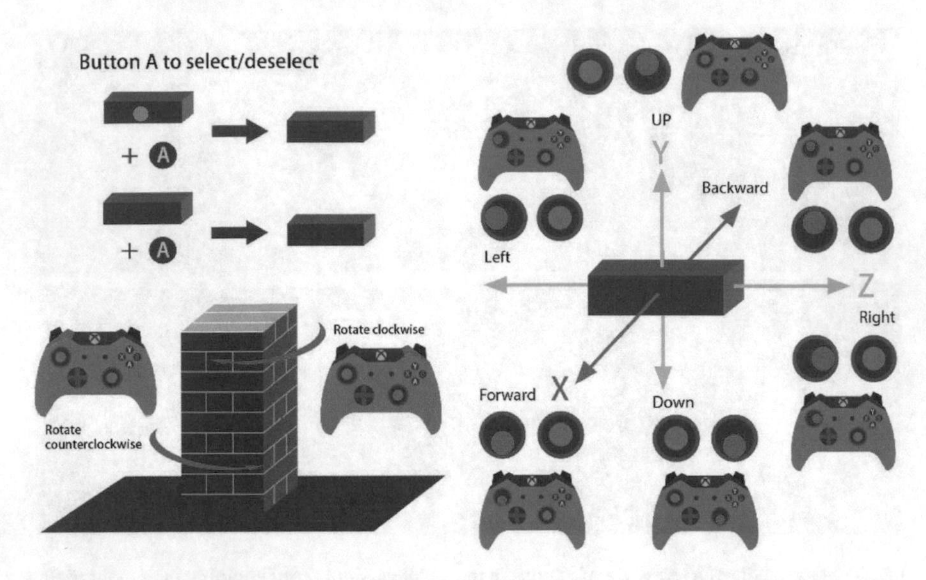

Fig. 10 The possible actions in the large display version

- The time between the selection of the block and when the block was deselected (Moving time).
- The result of each game (the total number of turns and each players' number of turns when the game ended, who won/lose/tied the game).

In addition, we collected the players' subjective feedback after gameplay.

3.2 Results and Discussion

Figure 12 summarizes the results of this study. The two pairs who used the VR platform made more moves and took a longer time to select and move blocks, especially in the second trial. The mobile tablet pairs made the fewest moves and spent an equal amount of time on each trial. Finally, the large display pairs sat in the middle in terms of numbers of moves, while they spent significantly more time in the second trial than in the first, similar to the VR groups.

Based on the number of moves, it would appear that the VR version enabled better gameplay experience. Players were better able to assess the blocks' positions and the overall stability of the tower. They were able to add more blocks taken out from the tower to the top while ensuring the blocks did not fall. From their subjective feedback, participants commented that they liked the immersive experience of the VR HMD. Beyond immersion, they also enjoyed the flexible viewing angles that could be achieved with relative ease—e.g., by simply walking around and moving their heads slightly. In other words, players were able to use more of their body parts

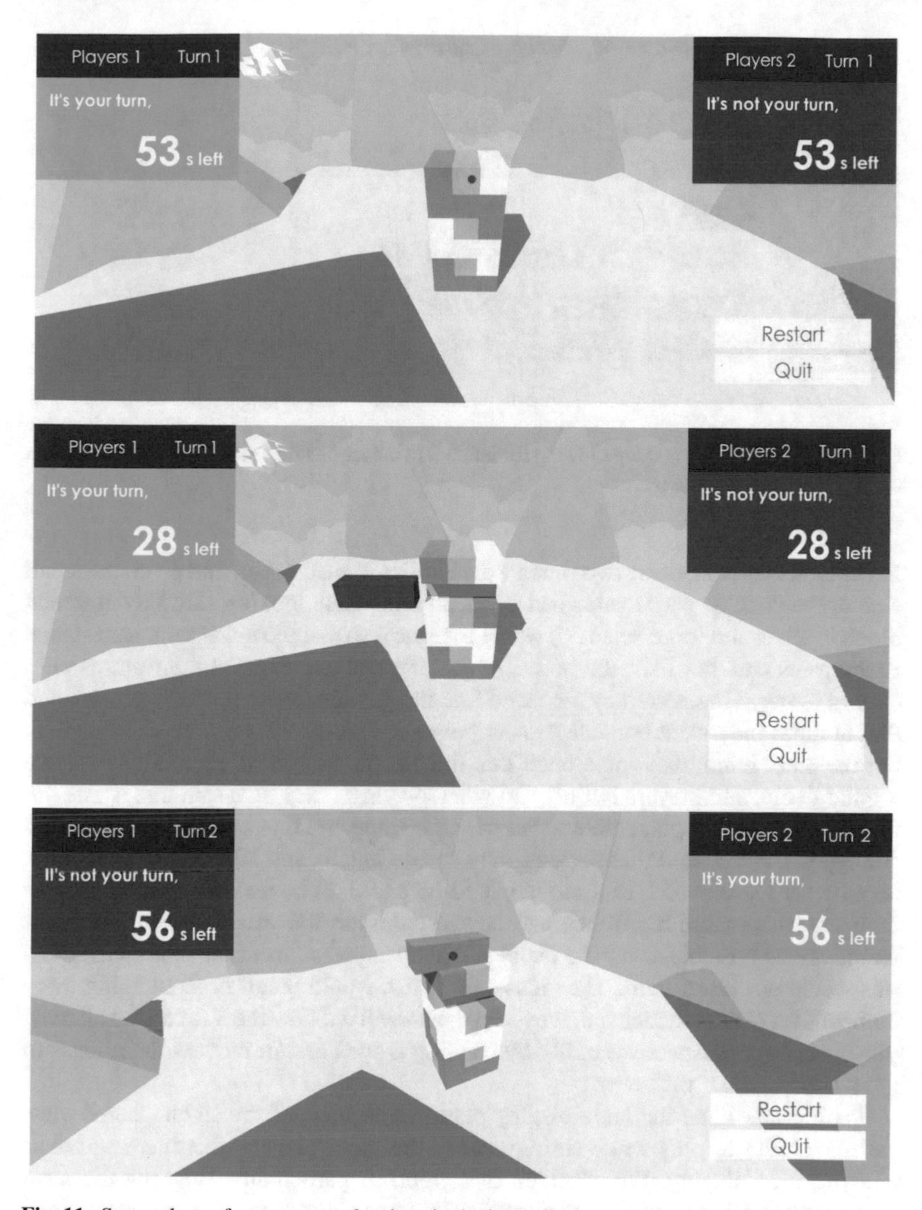

Fig. 11 Screenshots of a sequence of actions in the large display version: (Top) Selecting a block; (Middle) Moving a selected block out of the tower; and (Bottom) Placing it on top of the tower

and distribute the actions in ways that enabled them to see more details more easily and naturally.

Participants in both the tablet and large display versions had almost the same number of moves. However, they differed in terms of selection and movement times.

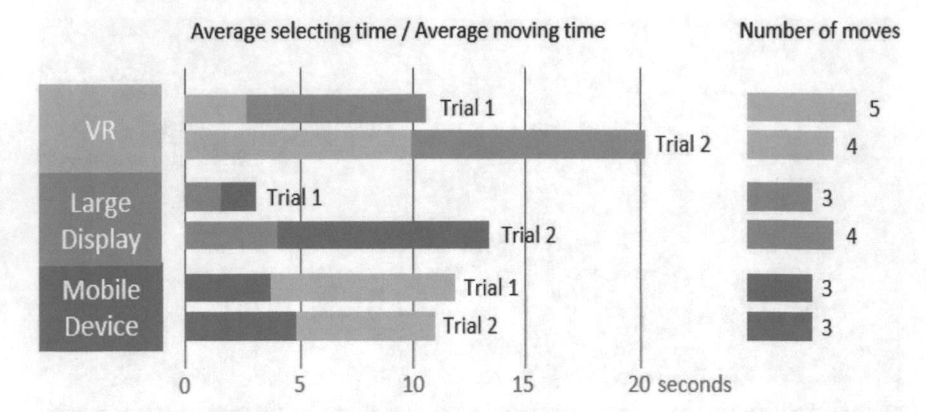

Fig. 12 The results of our study showing the number of moves and the average time taken for each move

In the tablet version, these two times were similar. We observed that given the small size of the display, participants had to spend more time looking carefully at which block to select and (once selected) where to place it so that the tower remained stable. Participants said they liked the variety of gestures and that they felt the gestures were natural to use. However, they also said that the gestures were not very precise and that at times they were not able to manipulate the blocks in the ways they wanted. On the other hand, they mentioned that the tablets allowed them to sit and move around freely, and players felt relaxed when playing. They also said that it was fun to see how the other player interacted with the game and find out their thoughts and strategies. Players said that at times they shared laughs and talked with each other about what the outcome of a move would be and did not feel that they wanted to compete with each other. When we showed them the VR version, they said that it would not be possible to see the facial and body expressions of the other player and this would not be as '*fun*'. They also said that it would result in them being more competitive. On the other hand, they said that they liked how the VR version allowed players change perspectives easily by walking around and/or moving their heads to get a closer look at the tower.

Participants liked the large display platform because of the single shared view and the ability to play while sitting next to the other player—this increased their enjoyment of playing. We observed this group of participants laughing the most and were actually trying to participate in the other player's turns as all the controls were taken over by one participant at a time. In comparison, in the VR and mobile versions, participants were able to view the blocks from different angles and think about their next move during the other person's turn, since the players had separate game displays. Like for the tablet platform, respondents desired greater flexibility in how they could change viewing perspectives—changing of perspectives was easier with the VR HMD where users only needed to turn or move their heads back-and-forth.

When playing the game wearing the VR HMD, participants appeared to be more focused on the game than the players in the other groups. This was another reason VR participants were able to make more moves, as they were competing and wanted to win. In fact, they barely talked with each other, though this was understandable as it was not possible for the VR participants to see each other. The other two groups (mobile and large screen) felt that they were playing the game more for pleasure than competition.

4 Summary and Future Work

In this project, we explored the usability issues and gameplay experiences of a multi-user casual, competitive game across three platforms: VR HMD, mobile tablets, and a large display. To this end, we developed a game based on the Jenga game but increased its complexity to make it more engaging and tested the game on 12 participants. We found that although all participants found the game engaging, they highlighted some positive affordances and features of each platform as well as elements that could be improved. In the future, we will use this feedback to improve the game's elements and expand its possibilities. For example, the current version can only be played by two players at a time, but this can be increased to a larger number of players. We intend to run more comparative studies to discover further ways of making these types of games enjoyable and engaging, specific to the various platforms available.

Acknowledgements We would like to thank the participants for their time and the reviewers whose comments and feedback have helped improve our chapter. This research was partially funded by the XJTLU Key Program Special Fund (KSF-A-03) and the XJTLU Research Development Fund.

References

Brouet, R., Blanch, R., & Cani, M. (2013). Understanding hand degrees of freedom and natural gestures for 3D interaction on tabletop. In *Human-Computer Interaction—INTERACT 2013* (Vol. 8117, pp. 297–314).

Hinckley, K., Pausch, R., Goble, J. C., & Kassell, N. F.(1994) A survey of design issues in spatial input. In *UIST '94 Proceedings of the 7th Annual ACM Symposium on User Interface Software and Technology*. Marina del Rey, CA, USA.

Kuittinen, J., Kultima, A., Niemelä, J., & Paavilainen, J. (2007). Casual games discussion. In *2007 Conference on Future Play*. New York, NY, USA.

Liang, J., & Green, M. (1994). JDCAD: A highly interactive 3D modeling system. *Computers and Graphics, 18*(4), 499–506.

Lindt, I., Ohlenburg, J., & Pankoke-Babatz, U. (2005). Designing cross media games. In *PerGames Workshop*, Pervasive 2005.

Lindt, I., Ohlenburg, J., Pankoke-Babatz, U., Prinz, W., & Ghellal, S. (2006) Combining multiple gaming interfaces in epidemic menace. In *CHI '06 Extended Abstracts on Human Factors in Computing Systems*. New York, NY, USA.

Oculus Rift | Oculus. (2016). [Online]. Available https://www3.oculus.com/en-us/rift/. Accessed October 31, 2016.

Pierce, J. S., Forsberg, A. S., Conway, M. J., Zeleznik, R. C. & Mine, M. R. (1997). Image plane interaction techniques in 3D immersive environments. In *Proceedings of the 1997 Symposium on Interactive 3D Graphics*. New York, NY, USA.

Rekimoto, J. (2002). SmartSkin: An infrastructure for freehand manipulation on interactive surfaces. In *Proceedings of the SIGCHI Conference on Human Factors in Computing Systems*. Minneapolis, MN, USA.

Tan, C. T., Leong, T. W., Shen, S., Dubravs, C., & Si, C. (2015). Exploring gameplay experiences on the Oculus Rift. In *2015 Annual Symposium on Computer-Human Interaction in Play*. New York, NY, USA.

Vive | Discover Virtual Reality Beyond Imagination. (2016). [Online]. Available https://www.vive.com/us/. Accessed October 31, 2016.

Vanacken, L., Grossman, T., & Coninx, K. (2007). 2007 IEEE symposium on 3D user interfaces. *Exploring the Effects of Environment Density and Target Visibility on Object Selection in 3D Virtual Environments*. Charlotte, NC, USA: IEEE.

Zhai, S. (1998). User performance in relation to 3D input device design. *SIGGRAPH Computer Graphics, 32,* 50–54.

An Inquiry Based Approach to Explore System Dynamics and Modeling Aspects of the Prey–Predator System

Tasos Hovardas and Zacharias Zacharia

Abstract This paper presents how a model-based inquiry approach can support the development of students' conceptual understanding about prey–predator systems as well as their cognitive skills. To do so, we developed a sequence of activities which involved three cycles of scientific inquiry. The first cycle revolved around the playing of a role-playing game about a prey–predator system, while the second cycle involved the development and running of a model (i.e., a simulation) of the prey–predator system on a computer. The third cycle followed the enactment of the first two cycles and entailed comparing the role-playing game to the computer simulation, to identify differences in the modelling process and system dynamics applied in each of the first two cycles and to identify and discuss nature-of-science related issues within the context of the prey–predator system. The complete approach was designed for upper secondary students, but with some adaptations it can also be used for lower secondary education or even at the two last grades of primary school.

Keywords Computer simulation · Model-based inquiry · Prey–predator system Role-playing games

1 Introduction

One of the major goals of science education is to enable students to transfer and apply the knowledge and skills acquired in prior learning experiences to new learning contexts (e.g., Georghiades 2000). Such a process would reveal the durability of students' understanding and learning gains from a prior learning experience, as well as their competence in applying the skills developed before in a new context.

T. Hovardas (✉) · Z. Zacharia
Research in Science and Technology Education Group, Department of Educational Sciences, University of Cyprus, P. O. Box 20537, 1678 Nicosia, Cyprus
e-mail: hovardas@ucy.ac.cy

Z. Zacharia
e-mail: zach@ucy.ac.cy

Y. Cai et al. (eds.), *VR, Simulations and Serious Games for Education*, Gaming Media and Social Effects, https://doi.org/10.1007/978-981-13-2844-2_9

Hence, educators aiming to evaluate the success of such transfers should include in their activity sequences at least two subsequent learning activities of varying context. This approach is perfectly suited to model-based inquiry, which involves a succession of model construction and refinement (Bamberger and Davis 2013). Through the enactment of multiple cycles of scientific inquiry, one after another, transfer will ideally be facilitated by means of metamodeling knowledge, which includes a thorough conceptualization of crucial model assumptions, the application of alternative models to examine the same phenomenon under study, and the revision and refinement of models to incorporate variations in the modelled phenomena (Fortus et al. 2015).

The aim of this paper is to present a model-based inquiry approach, which facilitates and encourages such transfers of knowledge and skills. The idea is to use a system dynamics example to show how acquired knowledge and skills can be further enhanced through subsequent model-based, inquiry-oriented learning cycles. Our suggested approach involves three inquiry-oriented learning cycles. The first cycle includes a learning activity sequence organized around a role-playing game of the prey–predator system. In this cycle, the students construct a concrete, physical game model in which they then participate. The second cycle includes a learning activity sequence that requires, among others, the construction of a simulation model of the prey–predator system on a computer. The third cycle follows after the first two cycles and entails comparing the game to the computer simulation. The idea for this third cycle is to examine nature-of-science and model (i.e., metamodeling knowledge) aspects and learning difficulties in addressing system dynamics in the prey–predator system. The example presented here is suitable for upper secondary education. However, after some adjustment, adapted modules might be used in lower secondary education or even in the two last grades of primary school.

2 The Prey–Predator System

The prey–predator system is mathematically described by the Lotka–Volterra equations (Lotka 1925; Volterra 1926):

$$Nt/dt = r * Nt - b * Nt * Pt \quad (1)$$

$$Pt/dt = d * Nt * Pt - c * Pt \quad (2)$$

where "N_t" and "P_t" stand for the prey and predator populations, respectively; "*r*" symbolizes the natural growth coefficient of prey; "*b*" is the predator efficiency; "*d*" corresponds to the biomass conversion efficiency; and "*c*" depicts the natural death coefficient of the predator (see Hovardas 2016 for a detailed presentation of the equations).

Population trends for both predator and prey undergo oscillations of standard width in time. This is counterintuitive to learners' initial ideas that are based on

Table 1 Number of individuals in the prey and predator populations in time

Unit of time (serial number)	No. of individuals in the prey population	No. of individuals in the predator population
1	10	4
2	6	8
3	2	12
4	10	4
…	…	…

linear reasoning, and which predict that predators will consume prey to exhaustion (e.g., Hokayem et al. 2015; Lefkaditou et al. 2014).

3 Role-Playing Game Model

To play the game, students are split in two groups (i.e., the "prey" and "predator" groups) and follow a series of four rules (adapted from Beals and Willard 2001 and adjusted to reflect the simulation requirements): (1) Each "predator" in the "predator" group is to capture one "prey" individual at each point in time, representing the way "prey" and "predator" populations "interact"; (2) each "prey" individual caught is to be converted into a "predator" individual; (3) all "predators" are to capture "prey"; and (4) if a "predator" cannot find any "prey" individual to catch, "it" is to be converted into "prey". Interaction between student groups is signaled whenever the instructor or a student claps their hands. "Prey" and "predator" individuals stand face to face and "interact". One particular game simulation that started with ten "prey" and four "predator" individuals, where students followed the rules listed, was observed and population data was recorded over time (see Table 1 for the "prey" and "predator" population sizes with time). We observed that the population sizes repeated every three-unit timespan (i.e. every three interactions between "prey" and "predator" populations).

A graphical representation of the game simulation is shown in Fig. 1. Instead of the predators consuming all the prey, the graph indicates that both populations remain in existence and fluctuate with time. Predators cannot consume all prey because consumption decreases prey availability and this introduces a limiting factor for the predator population (i.e., negative feedback loop for predation). As soon as the predator population starts to decrease due to limited prey abundance, the prey population begins to increase. The same sequence of events will be repeated since the number of predators will then increase due to increased prey availability, with predation constituting the limiting factor for the prey population. Prey and predator numbers present oscillations of constant width over time which emerge even if the system starts with different initial population sizes.

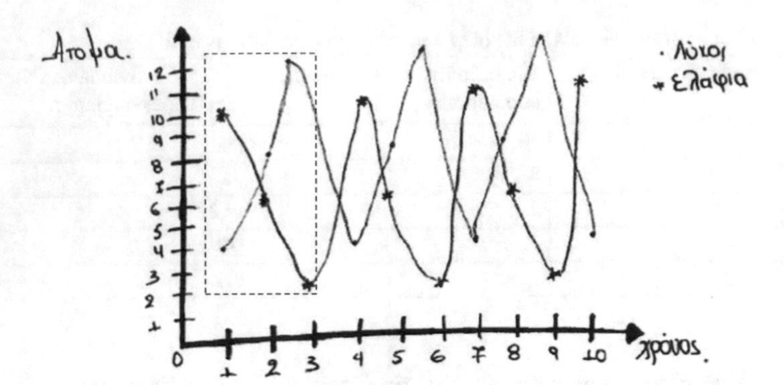

Fig. 1 Graph depicting prey (stars) and predator (dots) population trends. The dashed rectangle depicts the area of the figure where prey and predator population trends might be misinterpreted as being "inversely proportional"

This particular game is included in an inquiry-oriented activity sequence, which is presented in Table 2 (see Table 2 for: learning activities listed along with time needed to undertake each activity, class arrangement, and learning products constructed by students in each activity). In the context of this activity sequence, students first write down their initial ideas by responding to a hypothetical scenario involving a prey and predator population (e.g., hares and foxes or deer and wolves) in a hypothetical biotope. The basic assumptions for this hypothetical scenario are: (1) no individual exits the forest and that (2) no other individual enters, as well as that (3) there are no other prey or predator species present.

After students respond to this introductory scenario, the teacher presents them with scientific data that reveal the coexistence of prey and predator populations in various biotopes. This coexistence, after all, is a basic requirement for all trophic relations in food webs to be sustained. Students then play the game to simulate the prey–predator system by following the rules of the game, during which the numbers of "prey" and "predator" populations are recorded. Post-game, students construct graphs to observe population trends, then compare their interpretations with their initial ideas. Next, a whole-class discussion underlines the similarities and differences between students' graph interpretations and their initial ideas. Finally, the activity sequence concludes with the consideration of a new scenario set by the teacher (e.g., play the game with different initial population sizes, including a version of shrinking biotope; adding a hunter in the forest; adding a second prey species; and so on). The students predict population trends under the new scenario and attempt to revise the rules of the game to address the new learning context. Model-based inquiry can continue beyond the learning activities given in Table 2, based on available time, with the teacher providing feedback and students following the new rules to simulate the revised model and reflect on the new role-playing game.

Table 2 Learning activity sequence for the role-playing game of the prey–predator system

Learning activity (serial number)	Time to complete; class arrangement	Learning product
Eliciting initial ideas (1)	10 min; individual	Text addressing an introductory scenario
Comparing initial ideas with scientific data (2)	10 min; individual	Text focusing on the comparison at task
Playing the game (3)	20 min; group	Table with numbers of prey and predator individuals
Constructing a graph (4)	20 min; group	Graph presenting prey and predator population trends
Interpreting the graph (5)	10 min; group	Text focusing on the interpretation of the graph
Comparing graph interpretation with initial ideas (6)	10 min; group	Text focusing on the comparison at task
Discussing the comparison of graphs with initial ideas (7)	15 min; whole-class	Table which includes similarities and differences
Predicting prey and predator population sizes in a new learning context (8)	10 min; individual	Text addressing the new scenario
Revising the game to address the new scenario (9)	15 min; individual	Text which includes revised rules for the game

Class arrangement refers to students working each on his/her own (individual), in groups (group), or all together (whole-class)

4 Computer Simulation

The computer simulation of the prey–predator system can be operationalized by means of a dynamic feedback model software like STELLA. An example of the model to be constructed is given in Fig. 2. In the structure, we used the model interface provided by STELLA to transfer the products included in the Lotka–Volterra equations (“gains N” $= r*N_t$; “losses N” $= b*N_t*P_t$; “gains P” $= d*N_t*P_t$; “losses P” $= c*P_t$; products are shown by arrows connecting flows with stocks or flows with coefficients; see Hovardas 2016 for a detailed description of the modeling requirements of the prey–predator system in STELLA).

After inserting equations and values for all variables, the graph we obtain in STELLA is shown in Fig. 3. Curves resemble closely the ones derived from the game simulation (see Fig. 1) and show that population sizes of prey and predator undergo fluctuations of constant magnitude over time.

The learning activity sequence to be followed in the case of the computer simulation is presented in Table 3. Students first construct a computer simulation model of the prey–predator system by using the stocks, flows and coefficients provided in the model interface of STELLA. Products describing equations are indicated by arrows. After students enter values for all variables in the equation interface, they proceed to

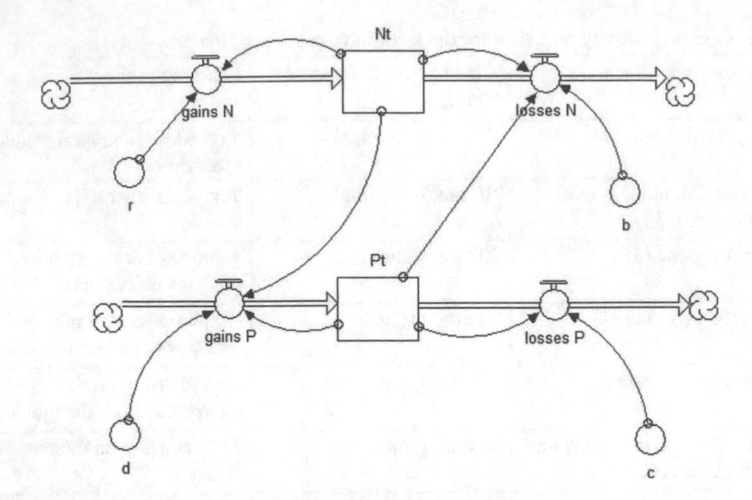

Fig. 2 The computer model of the prey–predator system constructed by STELLA

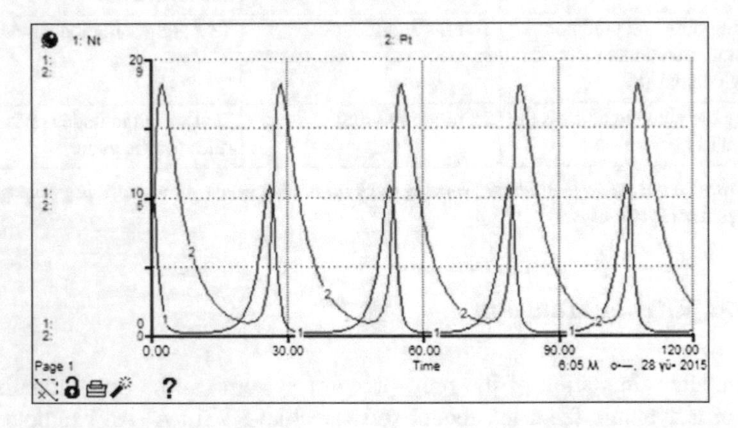

Fig. 3 Graph depicting prey (curve 1) and predator (curve 2) population trends in STELLA. Values for variables: $N_t = 10$, $P_t = 4$, $b = 0.2$, $d = 0.1$, $r = 0.5$, $c = 0.2$. To display the graph, we used the Runge–Kutta–Fehlberg integration method

simulate the prey–predator system, construct a graph, and interpret the graph. The next learning activity involves the construction of a scatterplot which is interpreted immediately after. Students then explore model behavior for different initial population sizes, predict population trends to respond to a new learning challenge, and revise their models to address this new challenge.

Table 3 Learning activity sequence for the computer simulation model of the prey–predator system

Learning activity (serial number)	Time to complete; class arrangement	Learning product
Constructing a model of the prey–predator system (1)	20 min; individual	Structural model of the prey–predator system
Inserting equations of population dynamics (2)	10 min; individual	Equations describing the prey–predator system
Simulating the prey–predator system (3)	10 min; individual	Graph depicting prey and predator population trends
Interpreting the graph (4)	10 min; whole-class	Text which focuses on the interpretation of the graph
Constructing a scatterplot with prey and predator populations (5)	10 min; individual	Scatterplot
Interpreting the scatterplot (6)	20 min; whole-class	Text which focuses on the interpretation of the scatterplot
Exploring model behaviour for different initial population sizes (7)	15 min; group	Scatterplot with multiple curves
Predicting prey and predator population sizes in a new learning context (8)	10 min; group	Text presenting a new scenario
Revising the model to address the new scenario (9)	15 min; group	Revised model

Class arrangement refers to students working each on his/her own (individual), in groups (group), or all together (whole-class)

5 Comparison Between the Role-Playing Game Model and the Computer Simulation Model

The final learning activity sequence of our approach involves a comparison between the role-playing game model and the computer simulation model of the prey–predator system. The first aspect that must be noted is the aligning of the game rules to the basic requirements of the mathematically depicted system. The predominant metamodeling feature shared by both models (i.e., the game model and the computer model) will then be discussed, namely that prey and predator populations comprise a common biomass pool. Finally, we will elaborate on the potential for misinterpretations of population dynamics in the graph of the game model which represents a crucial difference between the two modeling variants of the prey–predator system.

The rules which the game model is based on align the game with the equations describing the prey–predator system. Namely: Rule 1 (each "predator" in the "predator" group captures one "prey" individual at each point in time) refers to the fact that there is a threshold in predator efficiency; rule 2 (each "prey" individual caught is converted into a "predator" individual) corresponds to the conversion of prey biomass into predator biomass through predation; rule 3 (all "predators" capture "prey") concerns the main effect of predation, namely the fact that the predator population benefits from this trophic relation as long as there is prey available; and rule 4 (if a "predator" cannot find any "prey" individual to catch, it is converted into "prey") pertains to prey availability as a limiting factor for the predator population.

However, rule alignment does not result in a perfect match between the two simulations. For instance, the biomass conversion in the game simulation evolves differently compared to the biomass conversion in the computer simulation. These different operationalizations might be reflected in the highest and lowest values of the population curves (i.e., extreme points in population fluctuations). Comparing the maxima and minima of population sizes in the graph constructed from data generated by the game simulation (Fig. 1) and the graph constructed from data generated by the computer simulation (Fig. 3) shows that different extreme population fluctuation values are derived from the two simulations.

Another aspect linked to biomass conversion is the fact that it could lead us to the timeline encountered in both simulations and the time unit differences between the game and the computer simulation. As the game simulation involves a simplistic reworking of the differential equations of the prey–predator system, it requires a one-to-one conversion of prey individuals into predator individuals and vice versa. The computer simulation, on the other hand, is capable of much more detailed and intensive mathematical analyses. Differences between the two types of simulations highlight the need to fine-tune modelling decisions (i.e., metamodelling knowledge) according to the potential of each simulation and its limits of validity. At the same time, these differences underline the pedagogical value of offering multiple options for simulating the prey–predator system.

A core aspect that both simulations share is the constant width of oscillations which indicates that both populations were conceived as a common biomass pool. This similarity is the basic metamodeling aspect shared by both the game and the computer simulation. In the game model, this fixed common biomass pool is shown through the unchanging total number of individuals in prey and predator populations at each point in time (i.e., 14 individuals in our example, across all units of time). The computer simulation offers another way to visualize the common biomass pool, through a scatterplot of prey plotted against predator population size (Fig. 4). The cyclical shape formed in this scatterplot reflects the standard distancc between maxima and minima for both populations. The higher the number of individuals in the system (e.g., the higher the biomass in the common pool of both populations), the greater the radius of the cycle.

The game model succeeds in reproducing the constant width of population oscillations; however, it fails in reflecting the interrelation between the predator and prey populations that can be seen in the scatterplot derived from the computer simulation. This comparison reveals the oversimplifications often inherent in some modelling options (i.e., the game simulation cannot offer a perfect reproduction of all requirements of the computer simulation). Further, it showcases the limits of validity of a modelling strategy compared to another modelling strategy (i.e., the limits of validity of the game simulation as compared to the computer simulation) of the same system (i.e., the prey–predator system).

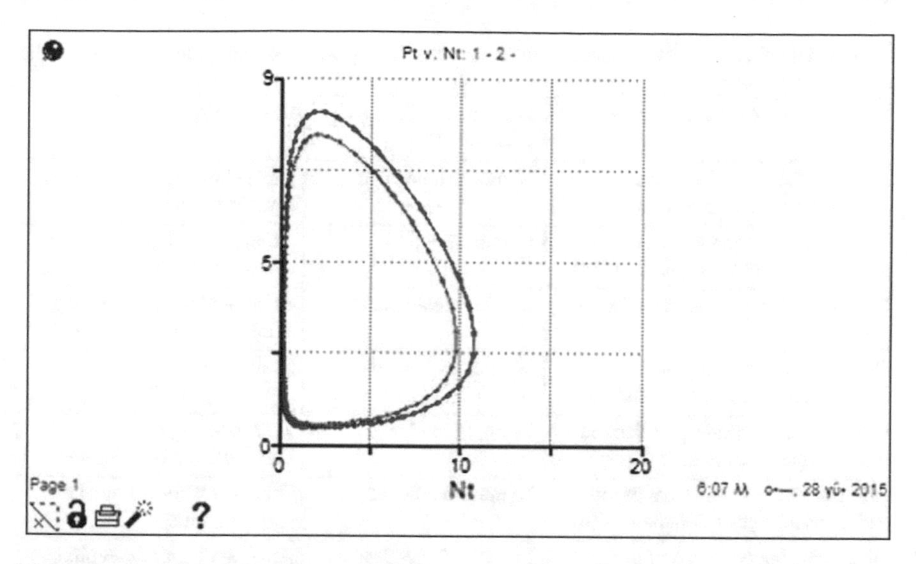

Fig. 4 Scatterplot depicting prey (Axis X) and predator (Axis Y) population trends in STELLA. Outer cycle: initial population sizes of 10 prey and 4 predator individuals; inner cycle: initial population sizes of 9 prey and 4 predator individuals

Another important aspect which emerged in the initial implementation of the game model (see Hovardas 2016) was that students who concentrated on the area outlined by maxima and minima of curves in graphs (see the dashed rectangle in Fig. 1) were quite likely to describe prey and predator population trends as being "inversely proportional". This represents a serious misinterpretation of system dynamics and a regression to linear thinking. Any variant of proportionality between prey and predator population sizes can only be justified by the plotting of populations against each other. Axis X in Fig. 1, however, depicts units of time.

These points to note have been integrated into the learning activity sequence that we suggest for comparing the game and the computer models of the prey-predator system (Table 4). The sequence starts with an elaboration of the game simulation rules according to mathematical equations and moves on to comparing the graphs derived from the game and computer simulations. The process also includes time graphs as well as scatterplots to depict the prey and the predator populations.

6 Implications for Future Research

The first implementation of our approach revealed quite satisfactory results in terms of learning gains (Hovardas 2016). However, more research is needed to explore the full potential of our method for developing a learning progression. Future studies should highlight how metamodeling knowledge might foster metacogni-

Table 4 Learning activity sequence for comparing the game and computer models of the prey–predator system

Learning activity (serial number)	Time to complete; class arrangement	Learning product
Aligning the rules of the game simulation with equations (1)	15 min; whole-class	Table showing how rules and equations match
Comparing the time graphs of the game and computer simulations (2)	10 min; group	Table presenting similarities and differences
Discussing the comparison of graphs (3)	15 min; whole-class	Annotated table presenting similarities and differences
Constructing a scatterplot for data derived from the game simulation (4)	10 min; individual	Scatterplot
Interpreting the scatterplot derived from the game simulation (5)	5 min; individual	Text focusing on the interpretation of the graph
Comparing the scatterplots of the game and computer simulations (6)	10 min; individual	Table presenting similarities and differences
Discussing the comparison of scatterplots (7)	15 min; whole-class	Annotated table presenting similarities and differences

Class arrangement refers to students working each on his/her own (individual), in groups (group), or all together (whole-class)

tive awareness of the prey–predator system. Since metacognition catalyzes transfer (Georghiades 2000; Kuhn and Dean 2004; Psycharis et al. 2014), metamodeling aspects of the prey–predator system might prove most insightful in addressing model revision and refinement in subsequent cycles of model-based inquiry. In this regard, metamodeling knowledge acquired in one learning context can then be applied to address a novel learning challenge.

The conceptualization of the prey–predator system as a common biomass pool is its primary metamodeling feature, which has been represented by the constant sum of individuals in both populations at each unit of time in the game simulation and by the varying radiuses of the cycles in the scatterplot depicting prey and predator population trends in STELLA. To adequately address new learning contexts (e.g. different initial population sizes, introduction of hunting in the hypothetical forest, introduction of a second prey species, and so forth) and revise their initial models, students must take into account this metamodeling feature and employ it in their learning activities. Future research should investigate if subsequent learning activity sequences in model-based inquiry might catalyze transfer through the facilitation of metamodeling knowledge. The time needed to undertake each activity, class arrangement, and learning products should also be revisited.

References

Bamberger, Y. M., & Davis, E. A. (2013). Middle-school science students' scientific modeling performances across content areas and within a learning progression. *International Journal of Science Education, 35,* 213–238. https://doi.org/10.1080/09500693.2011.624133.

Beals, K., Willard, C. (2001). *Environmental detectives*. University of California, Berkeley: Lawrence hall of science.

Fortus, D., Shwartz, Y., & Rosenfeld, S. (2015). High school students' meta-modeling knowledge. *Research in Science Education, 46,* 787–810. https://doi.org/10.1007/s11165-015-9480-z.

Georghiades, P. (2000). Beyond conceptual change learning in science education: Focusing on transfer, durability and metacognition. *Educational Research, 42,* 119–139. https://doi.org/10.1080/001318800363773.

Hokayem, A., Ma, J., & Jim, H. (2015). A learning progression for feedback loop reasoning at lower elementary level. *Journal of Biological Education, 49,* 246–260. https://doi.org/10.1080/00219266.2014.943789.

Hovardas, T. (2016). A learning progression should address regression: Insights from developing non-linear reasoning in ecology. *Journal of Research in Science Teaching, 53,* 1447–1470. https://doi.org/10.1002/tea.21330.

Kuhn, D., & Dean, D. (2004). A bridge between cognitive psychology and educational practice. *Theory into Practice, 43,* 268–273. https://doi.org/10.1207/s15430421tip4304_4.

Lefkaditou, A., Korfiatis, K., & Hovardas, T. (2014). Contextualizing the teaching and learning of ecology: Historical and philosophical considerations. In M. Matthews (Ed.), *International handbook of research in history, philosophy and science teaching* (pp. 523–550). Dordrecht: Springer.

Lotka, A. J. (1925). *Elements of physical biology*. Baltimore, USA: Williams and Wilkins.

Psycharis, S., Botsari, E., Mantas, P., & Loukeris, D. (2014). The impact of the computational inquiry based experiment on metacognitive experiences, modelling indicators and learning performance. *Computers & Education, 72,* 90–99. https://doi.org/10.1016/j.compedu.2013.10.001.

Volterra, V. (1926). Fluctuations in the abundance of a species considered mathematically. *Nature, 118,* 558–560. https://doi.org/10.1038/118558a0.

Supermarket Route-Planning Game: A Serious Game for the Rehabilitation of Planning Executive Function of Children with ASD

Siti Faatihah Binte Mohd Taib, Yuzhe Zhang, Yiyu Cai and Tze Jui Goh

Abstract This chapter discusses game-assisted learning for children with autism spectrum disorder (ASD). A serious game is designed for intervention in children with ASD and other special needs for supermarket route-planning based on mobile devices. It was created with the aim of training executive functioning skills whilst monitoring progress through the data recorded and collected. The methodology is described to highlight the architecture of the game and content design of the ten levels in addition to three parent components.

Keywords Serious games · ASD · Game design

1 Introduction

Computer-assisted learning (CAL) has been useful in aiding the learning of children with autism spectrum disorder (ASD) (Goldsmith and LeBlanc 2004). CAL technology has also been reviewed as an effective tool in teaching new things (Grynszpan et al. 2013), and promises to be successful in areas where other treatments have failed (Williams et al. 2002). Game-assisted learning for children with ASD could also be helpful in the training of executive functioning skills. 'Executive function' is a term that includes mental tasks such as planning, impulse control, mental flexibility, and inhibition, to name a few (Hill 2004).

Planning is a complex, and dynamic operation in which a sequence of planned actions must be constantly monitored, re-evaluated, and updated. It has been reported that both children and adults with autism demonstrated impaired performance on planning tests such as the Tower of London (ToL) task. The ToL task requires indi-

S. F. B. Mohd Taib · Y. Zhang · Y. Cai (✉)
Nanyang Technological University, Singapore, Singapore
e-mail: myycai@ntu.edu.sg

T. J. Goh
Institue of Mental Health, Singapore, Singapore

Y. Cai et al. (eds.), *VR, Simulations and Serious Games for Education*, Gaming Media and Social Effects, https://doi.org/10.1007/978-981-13-2844-2_10

viduals to move disks using as few moves as possible, according to a specific set of rules, from a predetermined sequence on three different pegs to arrive at a goal state. Computerized ToL tasks have enabled the measurement of participants' planning and motor execution times (Hill 2004).

The advancement in technology has made computer games easily available to the public. It has been observed that most children, including those with ASD, show an affinity to computers (Ploog et al. 2012; Cai et al. 2013, 2017; Lu et al. 2017). In this case, it could be advantageous to use game-assisted learning to help children with ASD improve their executive functions.

Currently, classroom use of digital tablets as a teaching and learning tool has become more common worldwide (Campigotto et al. 2012). Many parents and teachers of students with ASD have favorable opinions on the use of assistive technology (Cumming and Strnadova 2012). While the results of using digital technology in classrooms internationally have been encouraging, the research in this area is still in its early stages (Cumming and Strnadova 2012). This is especially so with regards to the use of digital tablets for assistive learning for students with special needs.

Unity3D, a game development platform, was used to create this mobile game. Being a mobile game application, it is easily accessed from and played at home. The supermarket route-planning is a game aimed at training the planning and mental flexibility executive functions of primary school children with ASD using a natural setting such as a supermarket environment. This game focuses on route-planning and distractor tasks. The player is required to find the shortest route to collect all the items on the shopping list, with the distance covered in-game measured and scored. The real-time data collected will also be used to keep track of each individual child's progress.

The rest of this chapter is organized as follows. Section 2 elaborates on the methodology used in the design and development of the game application. Data collection and networks are discussed in Sect. 3. Section 4 concludes the chapter.

2 Supermarket Route-Planning Game

2.1 Game Content Design

Supermarkets provide an environment suitable for carrying out route-planning tasks. The supermarket game depicts an orthographic view of a boy with a shopping trolley. The player needs to collect all the items stated on the shopping list by clicking the items on the shelves as shown in Fig. 1. Once a food item is clicked, the boy with the trolley automatically navigates a path towards the item. Only when the boy has reached the selected food item's shelf, does the item move and scale down into his trolley, if it is an item found on the shopping list. If not, a pop-up message appears, notifying the player that the item is not on the list.

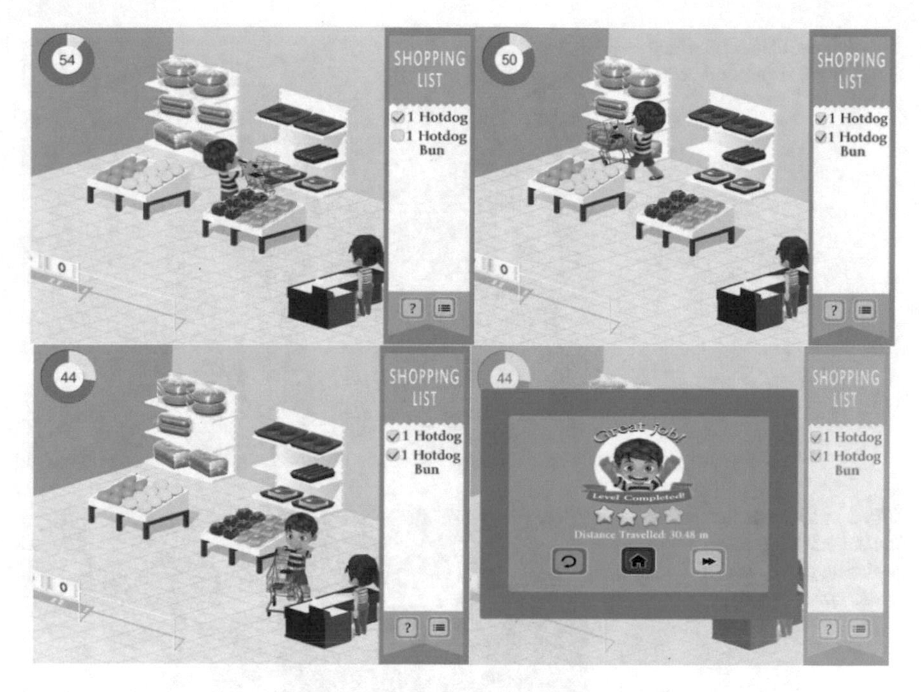

Fig. 1 Illustration of Level 1 and the level complete panel

Upon completion of a level, the total distance the player has travelled in that level is calculated and shown on the level complete panel for the player's view (Fig. 1 bottom-right). The player is then awarded stars based on their performance in that particular level. Three stars are awarded if the route that the player took is the shortest route. One and two stars are awarded according to predetermined ranges of distance travelled, which will be different for each level.

2.2 Level Design

The supermarket game consists of ten levels and three parent component levels (see Fig. 2). The first level serves as a tutorial and introduction to the game. The difficulty of the game increases with progression of levels. Each level has a time limit of a minute. In order to get at least one star in each level, players must complete the level within the time limit and meet the requirements of that particular level. A maximum of four stars can be obtained for each level, though the fourth star requires the use of the parent component.

As the player progresses in the game, the number of items on the shopping list increases. Level 1 has two items, Levels 2 and 3 have three items, Levels 4 and 5 have four items, and Levels 6–10 have five items on the shopping list. The layout of the

Fig. 2 Level User Interface (UI) for the supermarket game

Fig. 3 Layout of Levels 3 in the supermarket route planning game, with a distractor element

supermarket for Levels 1–4 is the same, both having a variety of ten food items. The difference between these four levels is the different locations of food items. The food items used are easily distinguishable, to minimise confusion in item identification for the user. Level 3 introduces a distractor element (see Fig. 3), i.e. a path blocker, which does not allow the player to pass through that particular path. Players have to recalibrate and plan out the shortest route.

The supermarket layout for Levels 5–10 is more complicated (see Fig. 4) than in Levels 1–4. There are 18 types of food items present in Levels 5–10. The distractor elements or path blockers come in the form of 'cleaning in progress' signs and stacks of soup cans, found in Levels 5, 8, 9 and 10 (see Fig. 4).

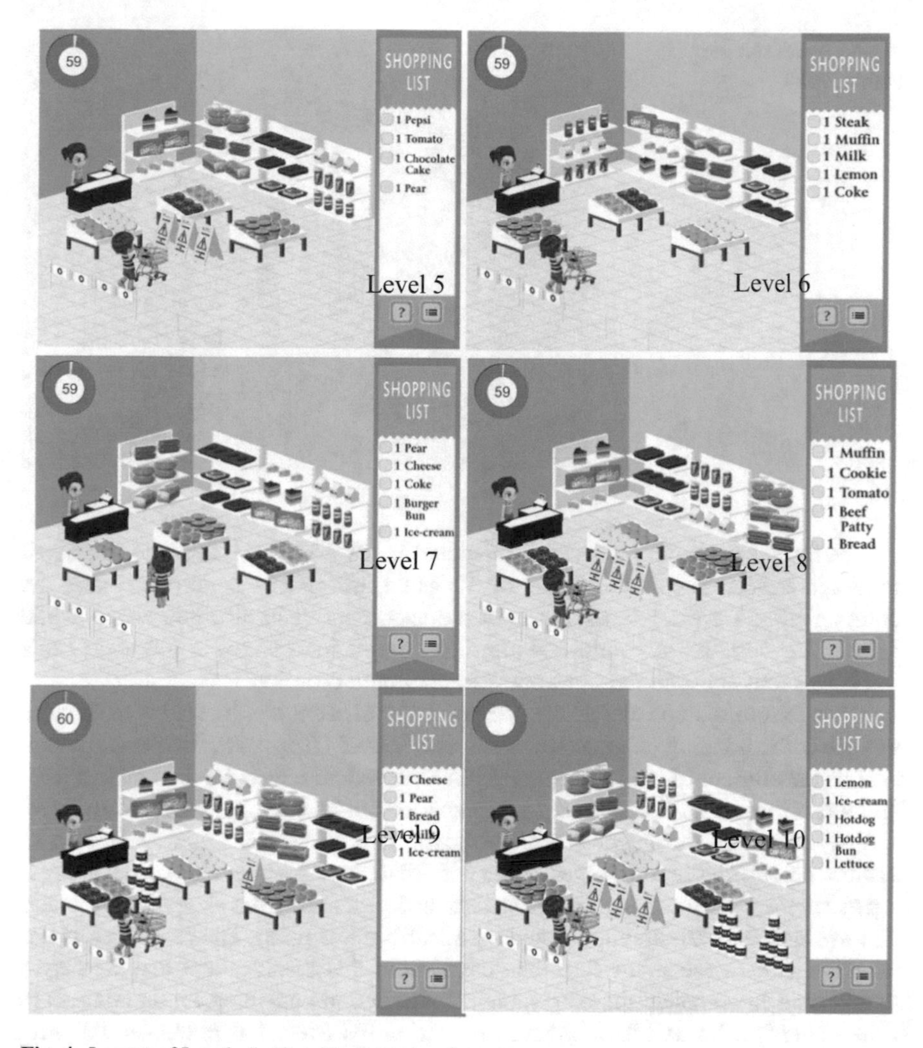

Fig. 4 Layout of Levels 5–10, with distractor elements present

2.3 Parent Component

A unique feature of this game is the inclusion of a parent component. It is introduced to encourage the participation of the player's parents, and to increase ecological validity and generalization of skills for the child in the real life environment. Before the unlocking of the three parent component levels, the maximum number of stars that can be obtained per level is three. In contrast, when a parent component level has been unlocked, the player now has the chance to achieve bonus fourth stars in that specific parent component's corresponding levels by executing the three-star performance for each level.

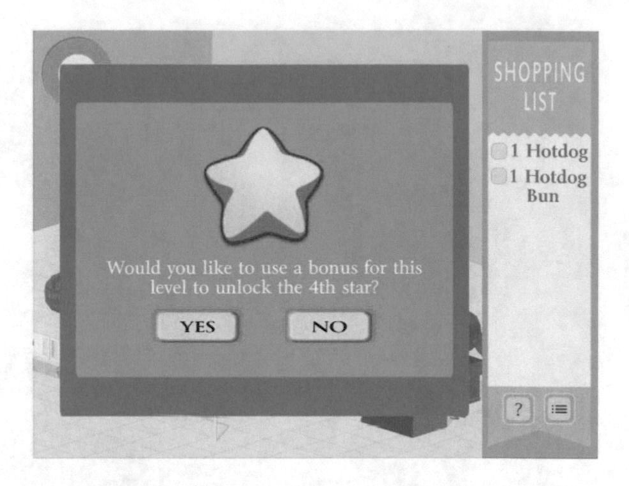

Fig. 5 Bonus star use pop-up panel

The three parent component levels, denoted as 'P1', 'P2' and 'P3', allows parents to assign tasks that befit the theme of the game. Level 'P1' is unlocked after the completion of Levels 1–3 with at least a one-star result in all three levels, while Levels 'P2' and 'P3' are unlocked after the completion of Levels 4–6 and Levels 7–10, respectively, with the same one-star minimum criterion.

Upon the completion of a parent component level, a counter located at the bottom of each corresponding parent component level's level UI appears. This statistic represents the number of bonus the player has achieved, to be used to complete a level and attain the fourth bonus star. The number of bonus attempts achieved in Level 'P1' would enable the player to achieve the fourth bonus star in Levels 1–3 only. To achieve the fourth star, the player would need reattempt those levels and select 'Yes' when a pop-up screen appears at the start of the level, asking the player if they would like to use their bonus attempt to achieve a fourth star (Fig. 5). The fourth bonus star criterion is the same as the three-star criterion of that level, except that the player would need to complete the corresponding parent component to obtain bonuses to achieve the fourth star. Likewise for Level 'P2' with Levels 4–6 and Level 'P3' with Levels 7–10.

Specifically, the parent component enables parents to help their child to practice the planning skills, by determining and selecting items for their child to help with on a grocery shopping list assignment. The number of items that can be assigned by the parents increases with each parent component in the game.

All three parent component levels are presented in a checklist format (see Fig. 6). Only the player's parent can certify the completion of each parent component level by entering their username and password when prompted. Each parent component level can be attempted as many times as desired or necessary.

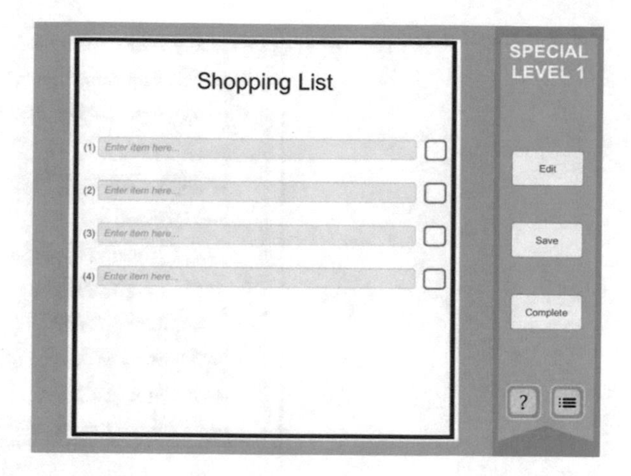

Fig. 6 Parent component 'P1' checklist

3 Data Collection and Networks

To track users' progress in real time, they are required to play the mobile game application with a network connection enabled. The generated data can thus be pushed onto a cloud server in real time rather than being saved on the local device. Each student with ASD must create their own account by signing up with a unique username and a 6-digit minimum password. The collected data is stored in the Firebase Realtime Database, a cloud-hosted database. The recorded data are as follow.

- Username of the player
- Date and time the level is played
- Level ID
- Status of level played (completed/times up/restart/quit)
- Score (corresponds with the number of stars achieved)
- Distance travelled by the player
- Time taken by the player to complete the level
- Number of correct items selected by the player
- Number of items on the shopping list
- Number of wrong items the player selected
- If the bonus (obtained from completing the parent component) is used.

With this data, we were able to monitor players' progress in their planning and mental flexibility executive function. An example of a data sample collected from this supermarket game is shown in Fig. 7.

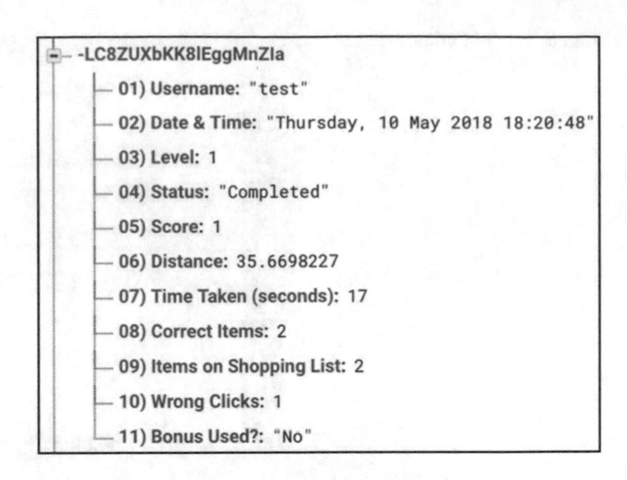

Fig. 7 Sample of data collected

4 Conclusion

The supermarket route-planning game is a mobile game application designed to help children with ASD train their planning skills through a game that is relatable to their everyday lives. The game also invites and encourages parents' involvement and participation via the use of encoded functions. This research looks into the benefits of game-assisted learning for children with ASD in improving their executive functioning skills.

However, as there were certain limitations in the experimental methods, clinical trials have yet to be conducted. In the future, the game could be adapted for mixed reality to produce an even more localised version for children with ASD.

Acknowledgements This project is supported by Rehabilitation Research Grant under the grant number RRG1/16002. The authors would like to thank all the people help in this project in one way or another. The chapter is dedicated to the memory of Mr. Husaini Md Rahim who passed away earlier this year.

References

Cai, Y., Chia, K. H., Thalmann, D., Kee, N., Zheng, J., & Thalmann, N. (2013). Design and development of a virtual dolphinarium for children with autism. *IEEE Transactions on Neural System and Rehabilitation Engineering, 21*(2), 208–217.

Cai, Y., Chiew, R., Nay, Z., Indhumathi, C., & Huang, L. (2017). Design and development of VR learning environments for children with ASD. *Interactive Learning Environments*. https://doi.org/10.1080/10494820.2017.1282877.

Campigotto, R., McEwen, R., & Epp, C. D. (2012). Especially social: Exploring the use of an iOS application in special needs class-rooms. *Computers and Education, 60,* 74–86.

Cumming, T. M., & Strnadova, I. (2012). The iPad as a pedagogical tool in special education: Promises and possibilities. *Special Education Perspectives, 21*, 34–46.

Goldsmith, T. R., & LeBlanc, L. A. (2004). Use of technology in interventions for children with autism. *Journal of Early Intensive Behavioural Intervention, 1*(2), 166–178.

Grynszpan, O., Weiss, P. L., Perez-Diaz, F., & Gal, E. (2013). Innovative technology-based interventions for autism spectrum disorders: A meta-analysis. *Autism*. https://doi.org/10.1177/1362361313476767.

Hill, E. L. (2004). Executive dysfunction in autism. *TRENDS in Cognitive Sciences, 8,* 1.

Lu, A., Chan, S., Cai, Y., Huang, L., Nay, Z., & Goei, S. L. (2017). Learning through VR gaming with virtual pink dolphins for children with ASD. *Interactive Learning Environments*. https://doi.org/10.1080/10494820.2017.1399149.

Ploog, B. O., Scharf, A., Nelson, D. & Brooks, P. J. (2012). Use of Computer-Assisted Technologies (CAT) to enhance social, communicative, and language development in children with autism spectrum disorders. Springer Science+Business Media, LLC.

Williams, C., Wright, B., Callaghan, G., & Coughlan, B. (2002). Do children with autism learn to read more readily by computer assisted instruction or traditional book methods? A pilot study. *SAGE Publications and the National Autistic Society, 6,* 71–91.

Virtual Reality Engine Disassembly Simulation with Natural Hand-Based Interaction

Yuan Xie, Yuzhe Zhang and Yiyu Cai

Abstract Virtual reality (VR) enhanced education has been widely studied recently and shown to be able to enhance students' learning motivation and efficiency. At the same time, rapid expansion of the aviation industry and technological innovation in aircrafts not only increase the demand for maintenance, repair, and overhaul (MRO) services, but also make MRO training costlier, especially in the case of aircraft engine MRO. In this chapter, a method is proposed to apply VR in aerospace engineering education and engine MRO training by simulating a virtual turbofan engine disassembly process. Natural hand-based gesturing solutions are investigated to enhance participants' interaction with the serious game.

Keywords Virtual reality · Serious games · Aircraft engine · Simulation

1 Introduction

Virtual reality (VR) is currently being widely studied in the education field and has shown promise in making learning experiences more interesting and amusing, thus improving learner motivation, engagement, and attention, especially when VR is combined with game elements (Virvou et al. 2005). VR is naturally suited for practical training because skills developed in a realistic virtual environment may transfer naturally to the real environment. This could be particularly useful in cases where the real environment is costly to provide, an example of which is the aircraft engine maintenance workshop (Mustafa and Carl 2015; Abulrub et al. 2011).

With the fast expansion of the aviation industry and advances in aircraft technology in recent decades, the aircraft maintenance, repair and operation (MRO) industry today faces not only a shortage of licensed aircraft mechanics and technicians but also the challenges of more complex updated tools and technologies. On one hand, older aircraft fleets generally require more frequent checks and maintenance, and on the

Y. Xie · Y. Zhang · Y. Cai (✉)
Nanyang Technological University, Singapore, Singapore
e-mail: myycai@ntu.edu.sg

Y. Cai et al. (eds.), *VR, Simulations and Serious Games for Education*, Gaming Media and Social Effects, https://doi.org/10.1007/978-981-13-2844-2_11

other hand, newer aircrafts are designed or manufactured with advanced technologies, which necessitate the investment in tools upgrade and maintenance techniques training by the airlines and MRO providers (Oliver Wyman 2015). As for the MRO market share by segment, engine MRO was reported to account for 42% in 2015 and is expected to grow continuously. On the other hand, safety issues associated with the aviation industry have attracted much public attention recently due to the increase in accidents, largely caused by human error in aircraft maintenance (Shanmugam and Robert 2015). Hence, the demand for better-educated engine engineers and trained technicians will continue to grow.

Thus far, aircraft technicians acquire almost 90% of their critical maintenance skills through on-the-job training (OJT), despite this approach being very costly for airlines and MRO providers. Many studies have been done in order to find new training systems to impart maintenance trainings to technicians. Some studies have indicated that computer-based training and virtual reality may be effective means for aircraft inspection and maintenance training (Sadasivan et al. 2004; Vora et al. 2002).

The purpose of this project is to propose a VR-enabled training method helping MRO technicians more efficiently trained on aircraft engine disassembly, through the use of interactive disassembly simulations with virtual aircraft engine models. This project also aimed to improve learning experience of aerospace engineering students on the topic of aircraft engine structure.

This chapter is organized in the following way: Sect. 2 discusses the modelling of aircraft engines. In the subsequent section, the design of the VR-enabled training simulation is illustrated, including its pedagogical structure design and interface design. Section 4 introduces the tools used during the project implementation. A user test that was conducted on the completed application is presented in Sect. 5. Finally, the project and potential future improvements to the VR engine are summarized in the last section.

2 Virtual Aircraft Engine Modelling

Due to the high cost, bulky volume, and heavy weight of real aircraft engines, they are not commonly found in university laboratories and thus cannot be examined and studied closely by aerospace engineering students. The same applies to MRO technicians who have to learn in the traditional way, i.e. from books or two-dimensional multimedia materials, before getting a chance to lay their hands on a real engine. Virtual engine models will thus provide more immersive and authentic learning experiences.

Aircraft engines have evolved into many different types to better serve a variety of purposes. From the turbofan engine's structure, it is clearly a fuel-efficient yet powerful engine and is thus economical for cruising. The Global Fleet and MRO Market Forecast reported that nearly 90% of the global fleet run today consists of jet aircraft including regional jets, narrow-body jets and wide-body jets (Oliver Wyman 2015). Most of these jet aircraft are equipped with turbofan engines. Therefore, the turbofan engine was selected for modelling as the interactive virtual engine in this project.

The model of the turbofan aircraft engine without cowling is shown in Fig. 1. Due to the complexity level of the actual engine, only its major and common functional components such as the low and high pressure compressor, combustion chamber, and inlet fan have been modelled (Rolls Royce 1996; Storm et al. 2007), as shown in Fig. 2. The engine model was built with triangular mesh to a high degree of precision. Statistics for some of the major components are listed in Table 1. The model has the typical appearance and structure of high-bypass turbofan engines like the Rolls Royce Trent 800 and the General Electrical GE90.

3 VR Engine Disassembly Simulation

The implementation of this project took into consideration both pedagogical concerns and interaction interface design, to be elaborated on in the following subsections.

Fig. 1 A side view of a virtual turbofan engine model

Table 1 Statistics of selected major model components

Component	Number of vertices	Number of polygons
Low pressure compressor	15,421	17,444
Inlet fan	5,804	11,728
Combustion chamber	10,475	18,674
Low pressure turbine	13,179	19,264

Fig. 2 Examples of major functional components

3.1 *Pedagogy Design*

The simulation procedure consisted of three parts (introduction, tutorial, and practice)—which correspond to students' cognitive processes of learning. In the introduction, users were presented with the side view and the exploded view of the virtual engine. At this stage, users observed the engine from different angles and hence formed a primary impression of the engine structure. Next, in the tutorial stage, users were guided by the system to disassemble the virtual engine according to the disassembly sequence at design level, as presented in the engine model's Illustrated Parts Catalog (IPC). The disassembly sequence could also be further customized to fit different disassembly manuals or learning purposes, allowing for maximized flexibility. The disassembly process was supported by various visual and audio aids to ease students' learning anxiety in the new environment and enrich learning modalities to fit different learning requirements or preferences. Lastly, in the practice stage, users practiced the disassembly sequence and reviewed the engine structure through system-generated questions about the location of specific components. By working through the questions, users digested the information at their own pace, further enhancing their knowledge retention over time (Fig. 3).

Fig. 3 Exploded view of virtual turbofan engine

3.2 Natural Hand-Based Interaction Interface Design

Three types of interaction interfaces were investigated in this project, namely a virtual button interface, ray casting from a fingertip, and ray casting from a camera. After each interface was tested and reviewed based on certain usability criteria, ray casting from a camera was selected because of its high selection precision, ease of learning, and immersive characteristics. Natural bare-hand gestures and interaction with the virtual engine were employed in the interface design to give users the illusion of touching, grasping, and manipulating the virtual objects directly with their own hands, increasing the degree of immersion in the experience (Poupyrev et al. 1998). Some examples of the gestures used are "grabbing" and "moving" the virtual engine components with a pinch gesture, confirming options with thumb-up gesture, and rotating the virtual engine by touching and pushing it with a bare hand. The use of this intuitive and natural interaction style was intended to greatly ease users' learning burden. Furthermore, realistic humanoid hands and forearms were utilized in the simulation, making the VR learning experience even more immersive and authentic (Fig. 4).

4 Project Implementation

Three major software applications and devices were used for project implementation: Oculus Rift, Unity 3D and Leap Motion. Oculus Rift is a head-mounted device (HMD) used for display and head motion tracking. Oculus Rift is a mature VR HMD with advanced display technology and a low-latency constellation tracking system

Fig. 4 Moving virtual engine components with natural hand gestures

that enables the user to feel actually present in the virtual environment. Unity 3D is a game development platform that was used to build the game's virtual environment because of its open support for VR devices and different CAD file formats as well as its highly-customizable developer setting. Leap Motion, a hand motion tracking device, was used to implement bare-hand gesture recognition.

5 User Test

A user test was carried out and a trial tutorial was piloted with this VR jet engine. The participants were students in the Computer-Aided Engineering (CAE) course at the School of Mechanical and Aerospace Engineering in Nanyang Technological University in Singapore, with most students being mechanical engineering majors without prior knowledge about aircraft engines. The students had previously learnt basic VR knowledge in the CAE course. To introduce the test, students were briefed on the background and operation procedures of the application before engaging in hands-on practice with the VR engine. A short online survey and quiz were done after the user test. The quiz questions were designed to measure what the students had learnt about basic engine structure from the application. The survey results revealed that most students thought the project helped them better understand turbofan engines. This finding was supported by the quiz results which showed that over 85% of students answered the questions correctly (Fig. 5).

Fig. 5 VR-enabled learning of turbofan engine

6 Conclusion

VR is a powerful tool that is revolutionizing the way people perceive knowledge. The aviation industry has been continuously developing and thus requires a larger supply of well-educated MRO engineers and technicians who are capable of working with current technologies. Efficient MRO technician training and engineer education have become a crucial challenge. To bridge this gap, this project proposed a VR-enabled pedagogical training application for engineering students and aircraft engine MRO technicians, aiming to help them better understand the turbofan engine's structure and disassembly process through interaction with a virtual turbofan engine via natural bare-hand gestures.

More work can be done to increase the value of this project. The current engine model does not possess a high degree of detail, and most components cannot be disassembled further into sub-assembly. More detailed engine components can be modelled so as to simulate second-stage disassembly. Moreover, the current engine model cannot run functionally. Either a mechanically connected engine model or an animation can be included to show the engine components working together with visible virtual air flow, highlighting the working principles of the engine. The educational value of this project will be significantly enhanced if these improvements are implemented in the future.

References

Abulrub, A. H. G., et al. (2011). Virtual reality in engineering education: The future of creative learning. In *2011 IEEE Global Engineering Education Conference (EDUCON)*.

Mustafa, H., & Carl, N. (2015). *The benefits of virtual reality in education—A comparison study*. Bachelor dissertation. University of Gothenburg.

Oliver Wyman. (2015). *2015–2025 global fleet and MRO market forecast*. [online] Oliverwyman.com. Available at: http://www.oliverwyman.com/our-expertise/insights/2015/oct/2015-2025-global-fleet—mro-market-forecast.html. Accessed April 18, 2017.

Poupyrev, I., et al. (1998). *The go-go interaction technique: Non-linear mapping for direct manipulation in VR*.

Rolls Royce. (1996). *The jet engine* (5th edn). England.

Sadasivan, S., et al. (2004). Technology to improve aviation safety: Recent effort at Clemson University. In *Proceedings of Safety across High-Consequence Industries Conference*.

Shanmugam, A., & Robert, T. P. (2015). Human Factors in Training for aircraft maintenance Technicians. In *Proceedings of 19th Triennial Congress of the IEA*.

Storm, R., Skor, M., et al. (2007). *Pushing the envelope: A NASA guide to engines*.

Virvou, M., Katsionis, G., & Manos, K. (2005). Combining software games with education: Evaluation of its educational effectiveness. *Educational Technology & Society, 8*(2), 54–65.

Vora, J., et al. (2002). Using virtual reality technology for aircraft visual inspection training: Presence and comparison studies. *Applied Ergonomics, 33*(6), 559–570.

Printed in the United States
By Bookmasters